AN
INTRODUCTION
TO THE
FINITE ELEMENT METHOD

THEORY, PROGRAMMING, AND APPLICATIONS

AN INTRODUCTION TO THE FINITE ELEMENT METHOD

THEORY, PROGRAMMING, AND APPLICATIONS

Erik G. Thompson

Professor of Civil Engineering
Colorado State University—Fort Collins

WILEY

JOHN WILEY & SONS, INC.

EXECUTIVE EDITOR	Bill Zobrist
PROJECT EDITOR	Jenny Welter
ASSISTANT EDITOR	Katie Mergen
MARKETING MANAGER	Ilse Wolfe
SENIOR PRODUCTION EDITOR	Valerie A. Vargas
SENIOR DESIGNER	Dawn Stanley
PRODUCTION SERVICES	Publication Services
COVER IMAGE	The bottom four images on the cover were created using MATLAB[®1] Graphics.

[1]MATLAB is a registered trademark of The MathWorks, Inc. For MATLAB product information, please contact

The MathWorks, Inc.
3 Apple Hill Drive
Natick, MA 01760-2098 USA
Tel: 508-647-7000
Fax: 508-647-7101
E-mail: info@mathworks.com
Web: www.mathworks.com

This book was set in 10/12 Times Roman by Publication Services and printed and bound by Hamilton Printing. The cover was printed by Phoenix Color, Corp.

This book is printed on acid free paper. ⊗

Library of Congress Cataloging in Publicaion Data:

ISBN 0-471-26753-8
ISBN 0471-45253-X

Printed in the United States of America
10 9 8 7 6 5 4 3 2 1

PREFACE

The purpose of this text is stated in its title: It is a balanced introduction to the theory, programming, and application of the finite element method. Thus there are four descriptive words.

Introduction. The intended audience for this book is first-year graduate students in engineering. It is also suitable for use in a senior-level course, where such courses are offered. Above all, it is a textbook. Each chapter has been written with students in mind. Where experience has proven a topic to be difficult or a possible trap for misunderstanding, care has been taken to give additional explanation. Matrix notation is used to help the students visualize the finite element matrices. Some knowledge of the use of partial differential equations to describe problems related to engineering analysis is assumed, as is a rudimentary knowledge of matrix algebra. However, knowledge of material that is unfamiliar to most first-year graduate students in engineering, such as the calculus of variations, is not assumed. Four appendixes have been included to describe material that can be incorporated into a course or simply left as reference material for the student to use.

Theory. The responsible use of finite element codes requires some understanding of the theoretical foundations upon which they are built. To impart such understanding is one of the goals of this text. Whenever a full treatment of the theory is beyond the scope of the text, an argument is presented that calls on students to use their intuition to grasp the basic concepts. This points the way for those who wish to explore the subject at a later time.

Programming. The availability of commercial software might lead one to believe that the teaching of finite element programming is unimportant. However, this is not the case. Programming presents an excellent platform with which to explain the numerical method, and it enhances understanding of the subject. Providing well-documented codes allows students a chance to test and experiment with the method as well as solve practical problems. For students who will use the finite element method in their research, having a code they can easily modify will be very valuable.

In this book, all codes are written in MATLAB script, which can be run on the student version of MATLAB. They have been written using explicit programming rather than MATLAB's implicit functions. Thus, they can be treated as pseudocodes to be transformed into FORTRAN or another language as desired. Each code is presented using a unique format: the left side of the page lists the code and the right side gives a detailed explanation of the code. In the exercises at the end of most chapters, there is a section on "Numerical Experiments and Code Development" with problems that encourage students to test and revise the codes provided, and to write auxiliary codes.

Applications. It is not until students have "hands-on experience" with the finite element method to solve complex engineering problems that they gain a full appreciation of the method and the confidence to use it on their own. For that reason the text covers a number of applications related to various fields of engineering. For each application, dimensional analysis is used to emphasize the physical meaning of the governing equations, the finite element approximations, and the relationship between the two. A number of challenging projects for students to undertake have been provided. Many of these projects encourage the use of the codes in new and creative ways, thus demonstrating the flexibility and the usefulness of the method.

It is hoped, therefore, that students who have completed a course based on this text will have a good understanding of finite element theory, the capability and confidence to modify existing codes and create new codes, and will have gained experience in using the finite element method to solve some interesting and nontrivial problems in engineering analysis.

Programming Codes and Data Files Available for Download. All codes, example data files, and auxiliary codes are available for download on the website for the book, located at www.wiley.com/college/thompson. Students should visit the Student Companion Site portion of the website to download the files.

Instructor Manual. The Instructor's Manual is available to instructors who have adopted the book for their course, and contains complete solutions, explanations, and suggestions. The Instructor Manual is available for download from the password-protected website at www.wiley.com/college/thompson. Visit the Instructor Companion Site portion of the website to register and request a password. Also available are example codes containing the suggested revisions associated with exercises in the book. These codes and corresponding data files are ready to run after downloading.

ACKNOWLEDGMENTS

The material in this text reflects the pursuits of many scholars as recorded in journals and books. I have draw deeply from this wellspring and wish to acknowledge this and express my sincere appreciation for their contributions. I also acknowledge my colleagues, both in academia and in industry, who have added to the contents of this book through many interesting and informative discussions over the years. To the students who used the notes on which this text is based and who were keen critics, I likewise expres my gratitude. It has been my good fortune that many of the above individuals are in two of the named groups and a few in all three.

I am very appreciative of those individuals at Wiley who accepted the manuscript and then guided me through the many stages that are necessary to prepare a textbook for publication—especially Wayne Anderson, Bill Zobrist, Bruce Spatz, Katie Mergen, Valerie Vargas, and Jenny Welter. Likewise, I extend thanks to Jan Fisher and Publication Services for their careful copyediting and production work. Very important to the completion of this text were the following reviewers who gave helpful suggestions:

Eniko T. Enikov, University of Arizona

Jiun-Shyan Chen, University of California, Los Angeles

Suresh Sitaraman, Georgia Institute of Technology

Scott Campbell, Ohio State University

James A. Sherwood, University of Massachusetts, Lowell

Theodor Krauthammer, Pennsylvania State University

Finally, I take this opportunity to thank Colorado State University and the Department of Civil Engineering for providing me the opportunity to teach the finite element method for over a quarter of a century. This was an exciting period in engineering analysis, and I consider myself fortuntate for being an eyewitness to it.

Erik Thompson
Colorado State University
Fort Collins
Fall semester 2003

CONTENTS

Appendix D   AUXILIARY CODES   323

BIBLIOGRAPHY   339

INDEX   341

INTRODUCTION

The finite element method[1] is a technique for obtaining approximate solutions to boundary value problems. To introduce the method, we present an elementary (some might say trivial) example of the deflection of a tightly stretched wire under a distributed load. This example is sufficient to (1) refresh the reader on some aspects of differential equations that will prove important for understanding approximate solutions, (2) introduce the concept of approximate solutions, and (3) actually define and illustrate the finite element method. This particular application was selected because the solution is easily visualized. However, any number of other applications could have been used, such as one-dimensional problems in heat transfer, porous flow, or electrostatics. The variables given here can be interpreted to correspond to one of these applications, and the reader is encouraged to do so if that is helpful.

1.1 GOVERNING EQUATION AND AN EXACT SOLUTION

Consider a tightly stretched wire as shown in Fig. 1.1. Under the proper circumstances, the deflection of the wire is accurately described by the solution of

$$T\frac{d^2y}{dx^2} + w(x) = 0.0 \tag{1.1}$$

where

T = tension in the wire $[F]$
y = deflection of the wire $[L]^2$
w = distributed load $[F/L]$

The bracketed terms are the dimensions of the variables, where F represents force and L represents length.

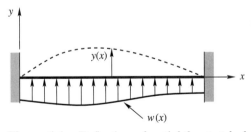

Figure 1.1. Deflection of a tightly stretched wire.

---

[1]To be abbreviated FEM throughout much of the text.
[2]Note that the symbol y represents a dependent variable and not a coordinate. Other symbols, such as u or v, could be used to eliminate any possible confusion; however, we have chosen to use y which conforms to the usage in many texts on ordinary differential equations. For two-dimensional problems in later chapters, y will indeed represent a coordinate.

The proper circumstances mentioned above are that T should be large enough and y small enough to ensure that T remains nearly constant during the deformation and that

$$\left(\frac{dy}{dx}\right)^2 \ll \left|\frac{dy}{dx}\right| \qquad (1.2)$$

Before we obtain an exact solution, it is necessary to define a particular loading. The loading selected is one that does not complicate the mathematics beyond that needed to fulfill the purposes stated above, and is shown in Fig. 1.2.

The governing equation now becomes

$$T\frac{d^2y}{dx^2} - 3W = 0 \quad \text{for} \quad 0 \le x \le L/2$$

$$T\frac{d^2y}{dx^2} - W = 0 \quad \text{for} \quad L/2 \le x \le L \qquad (1.3)$$

with boundary conditions

$$y(0) = y(L) = 0 \qquad (1.4)$$

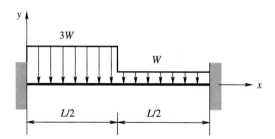

Figure 1.2. Loading on wire.

The loading is defined at all points but is discontinuous at $x = L/2$. The governing equation tells us that we should expect the same for the second derivative of y with respect to x. However, the first derivative must be continuous at $x = L/2$; otherwise the second derivative would not be defined at this point.

There are several approaches available to arrive at the exact solution. The one chosen here is to obtain the general solution for each of the two regions specified in Eq. 1.3, and require each one to have the same value for deflection and the same value for slope at $x = L/2$.

The general solutions for the two segments are

$$y(x) = \frac{W}{T}\left(\frac{3}{2}x^2 + C_1 x + C_2\right) \quad \text{for} \quad 0 \leq x \leq L/2$$

$$y(x) = \frac{W}{T}\left(\frac{1}{2}x^2 + C_3 x + C_4\right) \quad \text{for} \quad L/2 \leq x \leq L$$

$$(1.5)$$

The four constants of integration can be evaluated using the four conditions

$$y(x = 0) = 0$$

$$y(x = L) = 0$$

$$y\left(x = \frac{L}{2}-\right) = y\left(x = \frac{L}{2}+\right)$$

$$\frac{dy}{dx}\left(x = \frac{L}{2}-\right) = \frac{dy}{dx}\left(x = \frac{L}{2}+\right)$$

$$(1.6)$$

Evaluation of these constants gives us

$$y(x) = \frac{WL^2}{T}\left[\frac{3}{2}\left(\frac{x}{L}\right)^2 - \frac{5}{4}\left(\frac{x}{L}\right)\right] \quad \text{for} \quad 0 \leq x \leq L/2$$

$$y(x) = \frac{WL^2}{T}\left[\frac{1}{2}\left(\frac{x}{L}\right)^2 - \frac{1}{4}\left(\frac{x}{L}\right) - \frac{1}{4}\right] \quad \text{for} \quad L/2 \leq x \leq L$$

$$(1.7)$$

which is the exact solution to our governing equation. The plot of this function is shown in Fig. 1.3 at the end of the next section.

1.2 APPROXIMATIONS TO THE EXACT SOLUTION

We now consider how we would arrive at an approximate solution to our governing equation if the exact solution were not obtainable. Almost all methods for doing so use an approximating function with a finite number of degrees of freedom, such as

$$y(x) = (x)(x - L)\left[A_0 + A_1 x + A_2 x^2 + A_3 x^3\right]$$

$$(1.8)$$

or

$$y(x) = A_0 \sin\left(\frac{\pi x}{L}\right) + A_1 \sin\left(\frac{2\pi x}{L}\right) + A_2 \sin\left(\frac{3\pi x}{L}\right)$$

$$(1.9)$$

Here the degrees of freedom (the number of undetermined parameters) equal 4 in Eq. 1.8 and 3 in Eq. 1.9. Note that both approximations satisfy the boundary conditions independent of the parametric values. This is an important requirement for selecting approximating functions. Also worth noting is that the approximations can be made to converge to any continuous function by extending the series indicated. Such series are referred to as being *complete* or having the property of *completeness*, a topic covered in the mathematics of infinite series.

There are other series we could consider, including a finite element approximation. However, before we get to that, it is instructive to illustrate some of the methods used to obtain values for the parameters in these series. To do so, we will use the truncated polynomial series

$$y(x) = (x)(x - L)(A_0 + A_1 x) \tag{1.10}$$

Note that the dimensions of A_0 and A_1 must be $[L^{-1}]$ and $[L^{-2}]$, respectively. It is often best to write a series in a way that makes the parameters nondimensional, or at least in a way that gives them the same dimensions. However, to simplify the notation while retaining the physical dimensions of all quantities, the above form was chosen.

The exact solution satisfies the differential equation for all values of x. However, there are no values of A_0 and A_1 that would make our approximation capable of satisfying our equation at all points. This is easily seen by noting that the exact solution has a discontinuity in its second derivative, whereas all derivatives of our approximating function are continuous. Because there are no values for our parameters that will give us the exact solution, we must determine which values would be "best." This would not be a difficult task if we knew the exact solution. However, we are pretending not to know this, but only the differential equation that it must satisfy. We need, therefore, some technique for determining the best values for the parameters based only on the governing differential equation.

There are several ways to determine, or define, these "best" values, most of which are related to a function called the *residual*. This is the function obtained when the approximating function is substituted into the governing equation. For our case, this function is

$$R(x, A_0, A_1) = 2T A_0 + T(6x - 2L)A_1 - 3W \quad \text{for} \quad 0 \le x \le L/2$$

$$R(x, A_0, A_1) = 2T A_0 + T(6x - 2L)A_1 - W \quad \text{for} \quad L/2 \le x \le L \tag{1.11}$$

where we have indicated that the residual depends not only on x, but also on the values of the yet undetermined parameters A_0 and A_1. If it were possible to select values for A_0 and A_1 such that the residual would be zero for all values of x, then we would have the exact solution. However, as we have already noted, this is not possible. Hence, a compromise must be made. We now consider several well-known compromises.

1.2.1 Collocation. If we cannot make $R = 0$ at all points between 0 and L, we will make it zero at as many points as possible. Because we have only two undetermined parameters, we expect at most to be able to satisfy the equation at two points. We choose $x = L/4$ and $x = 3L/4$ as logical choices, and obtain

$$2A_0 - \frac{L}{2}A_1 = \frac{3W}{T}$$

$$2A_0 + \frac{5L}{2}A_1 = \frac{W}{T} \tag{1.12}$$

which gives us

$$A_0 = \frac{4W}{3T}$$

$$A_1 = -\frac{2W}{3LT}$$

(1.13)

Hence, our approximate solution by collocation is

$$\frac{y}{L} = \frac{WL}{T}\left[-\frac{2}{3}\left(\frac{x}{L}\right)^3 + 2\left(\frac{x}{L}\right)^2 - \frac{4}{3}\left(\frac{x}{L}\right)\right]$$

(1.14)

Figure 1.3, at the end of this section, compares this solution with the exact solution as well as the other approximate solutions that follow. The error between it and the exact solution, is shown in Fig. 1.4. Note that the maximum error is approximately 0.4%.

Two things are worth noting. First, the results are good because our approximation was good—a cubic that was able to duplicate our exact solution very closely. That might not be true for a more complex loading. In such a case, we would need to add more terms (parameters) in our approximating function. Second, our solution matches the exact solution at $x = L/4$ and $x = 3L/4$ (not easily seen from the graph, but it does). This is not because these points were the collocation points; rather, it is more of an accident peculiar to this particular problem. In general this will not be the case. We satisfy the *differential equation* at collocation points, not the *solution.*

1.2.2 Least Squares. Rather than making $R(x, A_0, A_1) = 0$ at two points, we might want to minimize its magnitude in some average sense along the entire length from 0 to L. One way of doing this is to minimize

$$J = \int_0^L R^2\, dx$$

(1.15)

Clearly, J is greater than zero except for the case where R is everywhere zero, i.e., the exact solution. Therefore, of all possible functions that satisfy our boundary conditions, the exact solution produces the lowest value of J. We know, because our approximation does not contain the exact solution, that it cannot produce this minimum value; thus, we settle for as low a value of J as it is able to produce. That is, we find values of A_0 and A_1 that give J a stationary value. Thus, we solve

$$\frac{\partial J}{\partial A_0} = 0 \quad \text{and} \quad \frac{\partial J}{\partial A_1} = 0$$

(1.16)

for A_0 and A_1.

Rather than integrate and then differentiate, it is easier to differentiate and then integrate; therefore, we write

$$\frac{\partial}{\partial A_i}\int_0^L R^2\, dx = \int_0^L 2R\frac{\partial R}{\partial A_i}\, dx = 0$$

(1.17)

Equations 1.11 give us R, from which we obtain

$$\frac{\partial R}{\partial A_0} = 2T \qquad \text{for} \quad 0 \le x \le L$$

$$\frac{\partial R}{\partial A_1} = T(6x - 2L) \quad \text{for} \quad 0 \le x \le L$$

(1.18)

From this we obtain the following two equations:

$$\frac{\partial J}{\partial A_0} = \int_0^{L/2} [2T][2TA_0 + T(6x - 2L)A_1 - 3W]\, dx$$

$$+ \int_{L/2}^L [2T][2TA_0 + T(6x - 2L)A_1 - W]\, dx = 0$$

(1.19)

$$\frac{\partial J}{\partial A_1} = \int_0^{L/2} [6T(6x - 2L)][2TA_0 + T(6x - 2L)A_1 + 3W]\, dx$$

$$+ \int_{L/2}^L [6T(6x - 2L)][2TA_0 + T(6x - 2L)A_1 + W]\, dx = 0$$

(1.20)

which gives, after integration and the solution of the resulting algebraic equations,

$$A_0 = \frac{5W}{4T}$$

$$A_1 = -\frac{W}{2LT}$$

(1.21)

Hence, our approximate solution found by the least squares method is

$$\frac{y}{L} = \frac{WL}{T}\left[-\frac{1}{2}\left(\frac{x}{L}\right)^3 + \frac{7}{4}\left(\frac{x}{L}\right)^2 - \frac{5}{4}\left(\frac{x}{L}\right) \right]$$

(1.22)

The solution is shown in Fig. 1.3 and its error in Fig. 1.4. It has a slightly higher maximum error than the function found by collocation, but is still less than 1%.

1.2.3 Galerkin's Method. Another method for reducing the residual over the entire length is to make its weighted average zero with respect to as many independent weighting functions as there are undetermined parameters. That is,

$$\int_0^L W_0 R(x, A_0, A_1, \ldots, A_n)\, dx = 0$$

$$\int_0^L W_1 R(x, A_0, A_1, \ldots, A_n)\, dx = 0$$

$$\vdots$$

$$\int_0^L W_n R(x, A_0, A_1, \ldots, A_n)\, dx = 0$$

(1.23)

The weighting functions must be independent in order for the resulting algebraic equations to be independent. Galerkin's method uses the independent terms of the approximating function as the weighting functions; thus, there are always as many weighting functions as there are independent parameters. For our example, the method gives us

$$\int_0^{L/2} \left[x^2 - Lx\right]\left[2TA_0 + T(6x - 2L)A_1 + 3W\right] dx$$

$$+ \int_{L/2}^{L} \left[x^2 - Lx\right]\left[2TA_0 + T(6x - 2L)A_1 + W\right] dx = 0 \qquad (1.24)$$

$$\int_0^{L/2} \left[x^3 - Lx^2\right]\left[2TA_0 + T(6x - 2L)A_1 + 3W\right] dx$$

$$+ \int_{L/2}^{L} \left[x^3 - Lx^2\right]\left[2TA_0 + T(6x - 2L)A_1 + W\right] dx = 0 \qquad (1.25)$$

Integration of these equations produces two algebraic equations in A_0 and A_1, the solution of which gives

$$A_0 = \frac{21W}{16T}$$

$$A_1 = -\frac{5W}{8TL} \qquad (1.26)$$

Thus, the approximation solution by Galerkin's method is

$$\left(\frac{y}{L}\right) = \frac{WL}{T}\left[-\frac{5}{8}\left(\frac{x}{L}\right)^3 + \frac{31}{16}\left(\frac{x}{L}\right)^2 - \frac{21}{16}\left(\frac{x}{L}\right)\right] \qquad (1.27)$$

1.2.4 The Ritz Method. For our tight wire problem it is possible to write an expression for the total potential energy of the system as a function of the deflection, $y(x)$. We know from the principles of mechanics that this scalar function attains a minimum value corresponding to the equilibrium position of the wire. That is, if the potential energy of the system is calculated using the exact solution of the governing differential equation, its value will be lower than that calculated when any other deflected shape is used for the calculation. Thus, as we did for the least squares method, we will judge the best values for A_0 and A_1 as those that give the smallest potential energy to the system, i.e., those values that produce a potential energy nearest to the potential energy that the exact solution has.

The potential energy of the system is given by

$$E = \int_0^L \left[\frac{1}{2}T\left(\frac{dy}{dx}\right)^2 - wy\right] dx \qquad (1.28)$$

for any loading $w = w(x)$ and any deflection $y = y(x)$.[3] Similar expressions exist for other applications.

---

[3]This can be shown using elementary mechanics. The first term in the integrand represents the potential energy of the distributed load, and the second term in the integrand represents the increase in potential energy (strain energy) in the wire due to the deflection.

We will shortly show that the minimization of such an expression is equivalent to satisfying the governing differential equation.

When an approximating function is substituted into the above integral, E becomes a function of the undetermined parameters used for the function; thus, to obtain a stationary value, we enforce

$$\frac{\partial E}{\partial A_i} = \frac{\partial}{\partial A_i} \int_0^L \left[\frac{1}{2} T(y')^2 - wy \right] dx$$

$$= \int_0^L \frac{\partial}{\partial A_i} \left[\frac{1}{2} T(y')^2 - wy \right] dx \qquad (1.29)$$

$$= \int_0^L \left[y'T \frac{\partial(y')}{\partial A_i} - w \frac{\partial(y)}{\partial A_i} \right] dx$$

For our two-parameter approximating function,

$$y = (x^2 - xL)A_0 + (x^3 - x^2L)A_1$$

$$y' = (2x - L)A_0 + (3x^2 - 2xL)A_1$$

$$\frac{\partial(y)}{\partial A_0} = (x^2 - xL)$$

$$\frac{\partial(y)}{\partial A_1} = (x^3 - x^2L)$$

$$\frac{\partial(y')}{\partial A_0} = (2x - L)$$ \qquad (1.30)

$$\frac{\partial(y')}{\partial A_1} = (3x^2 - 2xL)$$

$$w = \begin{cases} -3W & \text{for} \quad 0 \le x \le L/2 \\ -W & \text{for} \quad L/2 \le x \le L \end{cases}$$

Substitution of these functions into Eq. 1.29 creates, as with the previous methods, two linear algebraic equations for A_0 and A_1. Solving, we find that the resulting values of our two parameters are identical to the ones we obtained by Galerkin's method. There is a reason for this, which will soon be explained. The plots of this function and its error are the same as the plots for Galerkin's method in Figs. 1.3 and 1.4.

For many of the boundary value problems associated with engineering analysis, there is a corresponding scalar function such as we have used here. These scalar functions of functions[4] are referred to as *functionals*. The procedures for determining the functional corresponding to a given differential equation, or vice versa, are the subject of the calculus of variations. The principle that the solution of a given boundary value problem

---

[4]The function for E was a function of the function $y(x)$.

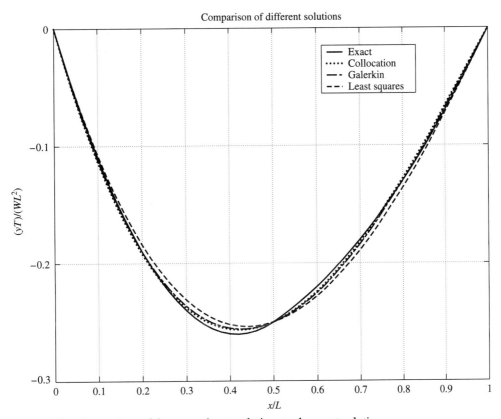

Figure 1.3.   Comparison of the approximate solutions to the exact solution.

corresponds to a stationary value of a functional is referred to as a *variational principle*. Finally, the concept of using a variational principle to judge the quality of an approximate solution is referred to as the Ritz method. We will spend considerable time exploring these ideas because they represent the foundations of the finite element method.

1.2.5 Results. In comparing the preceding methods, it is important not to place emphasis on which one has the smallest error. The errors shown depend more on the particular problem and the approximating function selected than on the method. All of the methods converge to the exact solution as the number of terms used in the approximation increase, provided that the series is complete. The question of convergence is important and will be discussed at the conclusion of this chapter.

For now, however, let us see how each of the methods fared. Figure 1.3 illustrates the deflection found for each method (the Ritz method is the same as Galerkin's). Figure 1.4 illustrates the difference between the exact solution and the approximate solutions. Again, you should understand that the closeness of these approximations to the exact solution is due to the fact that our approximating function, with only two independent parameters, can be made very close to the piecewise parabola that is the exact solution. In general, many more terms are necessary to obtain the accuracy needed in most engineering analyses.

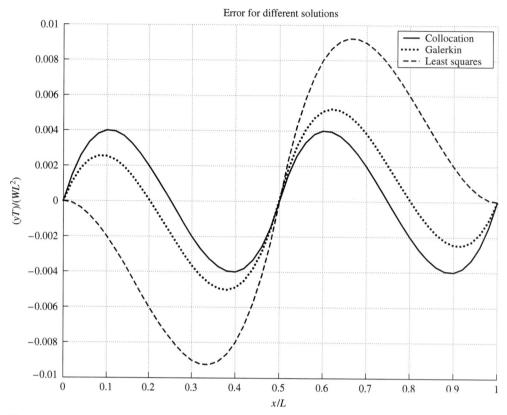

Figure 1.4.   Difference between the exact solution and the approximate solutions along the wire.

1.3 FINITE ELEMENT APPROXIMATIONS

There is a very important difference between the Ritz method and the other three methods illustrated in the previous section, and that difference allows us to introduce finite element approximations. Whereas the first three methods required that the second derivative of the approximating function exist, the Ritz method required only the existence of the first derivative. This weaker demand allows us to use piecewise linear functions as our approximating function. Such approximations are the foundations of the finite element method, several of which are illustrated in Fig. 1.5. You should note three characteristics of these functions, which we will discuss in much more detail in later chapters:

1. The functions are continuous, although their first derivatives (slopes) are not.
2. Each function is completely defined by its nodal point values; hence, these values serve as the undetermined parameters of the approximating function.
3. The segments, or elements, need not be of equal length.

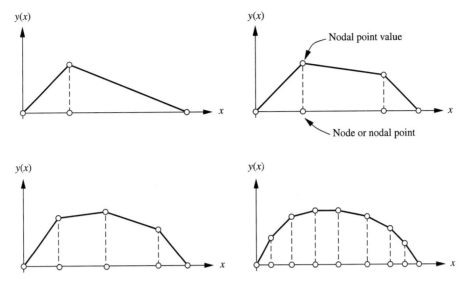

Figure 1.5. Finite element approximations.

Let us now use one such function to approximate the deflection of our tight wire. We choose the particular one illustrated in Fig. 1.6, given by

$$y(x) = \begin{cases} (3Y_2/L)x & \text{for} \quad 0 \le x \le L/3 \\ (2Y_2 - Y_3) + (3/L)(Y_3 - Y_2)x & \text{for} \quad L/3 \le x \le 2L/3 \\ 3Y_3 - (3Y_3/L)x & \text{for} \quad 2L/3 \le x \le L \end{cases} \tag{1.31}$$

The energy functional,

$$E = \int_0^L \left[-wy + 0.5T \left(\frac{dy}{dx} \right)^2 \right] dx \tag{1.32}$$

requires the use of the derivative of this function, which is

$$\frac{dy}{dx} = \begin{cases} 3Y_2/L & \text{for} \quad 0 \le x \le L/3 \\ (3/L)(Y_3 - Y_2) & \text{for} \quad L/3 \le x \le 2L/3 \\ -3Y_3/L & \text{for} \quad 2L/3 \le x \le L \end{cases} \tag{1.33}$$

Substitution of the function, its derivative, and the loading function into Eq. 1.32 gives us the potential energy as a function of the deflections, Y_2 and Y_3, which we indicate by writing

$$E = E(Y_2, Y_3) \tag{1.34}$$

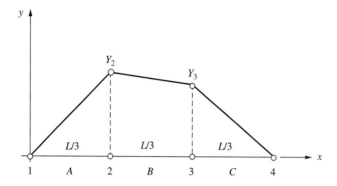

Figure 1.6. Finite element approximation to the deflection of the tight wire.

The Ritz method states that the best approximation to the exact solution that these two parameters can give are the two values that create the lowest value for E, that is, the value of E that is closest to the value obtained by the exact solution. Thus, we seek the two nodal values that will satisfy

$$\frac{\partial E}{\partial Y_2} = 0 \quad \text{and} \quad \frac{\partial E}{\partial Y_3} = 0 \tag{1.35}$$

As before, it would be easier to differentiate and then integrate, but this time we choose to integrate first and obtain the potential energy in terms of our two nodal values for displacement. This approach emphasizes that the potential energy is a quadratic function of the displacements; hence, it has one, and only one, stationary value.

A significant consequence of piecewise approximations is that the integration can be performed element by element and summed to obtain the total integral. This is the approach we now use.

The contribution to E from element A is

$$E_A = -\int_0^{L/3} \left[-3W\left[(3Y_2/L)x\right] \right] dx + \int_0^{L/3} \left[\frac{1}{2}T\left[3Y_2/L\right]^2 \right] dx \tag{1.36}$$

$$E_A = \left[\frac{WL}{2} \right] Y_2 + \left[\frac{3T}{2L} \right] Y_2^2 \tag{1.37}$$

The contribution to E from element B is

$$E_B = -\int_{L/3}^{L/2} \left[-3W\left[(2Y_2 - Y_3) + (3/L)(Y_3 - Y_2)x\right] \right] dx$$

$$- \int_{L/2}^{2L/3} \left[-W\left[(2Y_2 - Y_3) + (3/L)(Y_3 - Y_2)x\right] \right] dx$$

$$+ \int_{L/3}^{2L/3} \left[\frac{1}{2}T\left[(3/L)(Y_3 - Y_2)\right]^2 \right] dx \tag{1.38}$$

$$E_B = \left[\frac{5WL}{12}\right]Y_2 + \left[\frac{3WL}{12}\right]Y_3 + \left[\frac{3T}{2L}\right](Y_3 - Y_2)^2 \tag{1.39}$$

The contribution to E from element C is

$$E_C = -\int_{2L/3}^{L}\left[-W\left[3Y_3 - (3Y_3/L)x\right]\right]dx + \int_{0}^{L/3}\left[\frac{1}{2}T\left[-3Y_3/L\right]^2\right]dx \tag{1.40}$$

$$E_C = \left[\frac{WL}{6}\right]Y_3 + \left[\frac{3T}{2L}\right]Y_3^2 \tag{1.41}$$

The total energy is

$$E = E_A + E_B + E_C \tag{1.42}$$

$$E = \left[\frac{11WL}{12}\right]Y_2 + \left[\frac{5WL}{12}\right]Y_3 + \left(\frac{3T}{L}\right)\left(Y_2^2 - Y_2Y_3 + Y_3^2\right) \tag{1.43}$$

which is a quadratic function of Y_2 and Y_3. It thus has a single, stationary value where

$$\frac{\partial E}{\partial Y_2} = 0.0 \tag{1.44}$$

and

$$\frac{\partial E}{\partial Y_3} = 0.0 \tag{1.45}$$

Thus, we obtain

$$\left(\frac{11WL}{12}\right) + \left(\frac{6T}{L}\right)Y_2 - \left(\frac{3T}{L}\right)Y_3 = 0.0$$
$$\left(\frac{5WL}{12}\right) - \left(\frac{3T}{L}\right)Y_2 + \left(\frac{6T}{L}\right)Y_3 = 0.0 \tag{1.46}$$

or, in matrix notation,

$$\begin{bmatrix} \left(\dfrac{6T}{L}\right) & \left(\dfrac{-3T}{L}\right) \\[2ex] \left(\dfrac{-3T}{L}\right) & \left(\dfrac{6T}{L}\right) \end{bmatrix} \begin{Bmatrix} Y_2 \\[2ex] Y_3 \end{Bmatrix} = \begin{Bmatrix} -\dfrac{11WL}{12} \\[2ex] -\dfrac{5WL}{12} \end{Bmatrix} \tag{1.47}$$

The solution is

$$\left\{ \begin{array}{c} Y_2 \\ \\ Y_3 \end{array} \right\} = \left\{ \begin{array}{c} -\dfrac{WL^2}{4T} \\ \\ -\dfrac{7WL^2}{36T} \end{array} \right\} \tag{1.48}$$

This solution is shown in Fig. 1.7. The nodal point values in this case turn out to coincide with the exact solution. However, this will not happen under different circumstances (i.e., a different governing equation).[5]

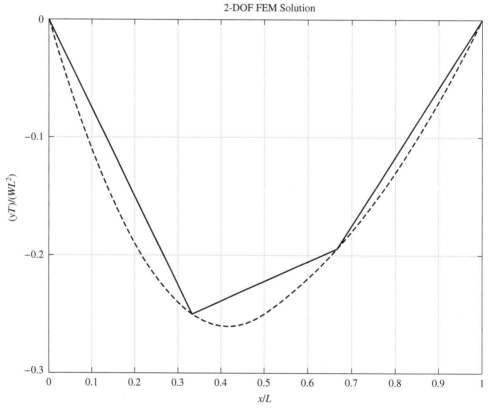

Figure 1.7. Finite element solution.

---

[5]The deflection of a tightly stretched wire under the action of concentrated loads is made up of piecewise linear functions, the same as our finite element approximation. Hence, our finite element approximations can be made to correspond to the exact solution simply by placing nodes at each point where there is a concentrated load. Furthermore, it can be shown that the exact solution for the deflection of a wire at any two points is the same for all statically equivalent loadings between these two points. These two facts are the reason for the remarkable results here. Proof of the latter of the two statements is an interesting exercise for students of applied mechanics.

You have probably noted that this solution does not compare well with those found by our two-term polynomial approximation. However, as more terms are added, the polynomial and trigonometric approximations become ever more difficult to evaluate, whereas the finite element solution simply repeats the same calculations, over and over, for each element. This advantage becomes pronounced for partial differential equations with irregular boundaries and nonconstant parameters.

1.4 BASIS FUNCTIONS

We have referred to our approximating functions as series. This is, of course, obvious for the polynomial and trigonometric approximations; however, finite element approximations do not, at first glance, appear to be a truncated series. Nevertheless, in much of the literature associated with the finite element method the approximation is treated as a series and written in the form

$$y(x) = \sum M_i(x)Y_i \tag{1.49}$$

When this is done, the functions $M_i(x)$ are called basis functions, with one such function associated with each node. A set of these functions for a four-node approximation is shown in Fig. 1.8.

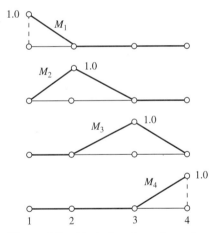

Figure 1.8. Basis functions for a four-node approximation.

Note that the magnitude of each function at its respective node is unity and that it is zero at all other nodes. Hence, their sum would have a value of unity at each node, and the function would simply be the straight line $y = 1.0$. If each basis function is multiplied by a constant A_i, then the sum of these new functions would be a piecewise linear function having a value at each node equal to the corresponding value of A_i for that node. As an example, if

$$A_1 = 0, \qquad A_2 = 4, \qquad A_3 = 3, \qquad A_4 = -1$$

then

$$y(x) = \sum A_i M_i(x) \tag{1.50}$$

is the function shown in Fig. 1.9.

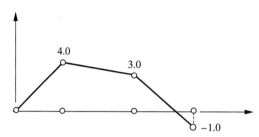

Figure 1.9. $\sum M_i Y_i$ for $\{Y\} = \{0, 4, 3, -1\}$.

1.5 WEAK FORMULATION OF PROBLEM

As we have seen, the use of a piecewise linear approximation depends on being able to express our problem in terms of the potential energy functional, Eq. 1.28. We also saw that for our polynomial approximation, the Ritz method produced the same results as Galerkin's method, and we noted that this was no accident. We now show how to obtain the integral formulation from the differential equation—that is, how to obtain the weaker formulation that requires only the existence of the first derivative from the stronger formulation that requires the existence of the second derivative. Both formulations are valid, and the solution to one is the solution to the other. Note specifically that what follows is not directly related to approximation theory, although we will certainly use the results for that purpose. That is, the original differential equation for our tight wire problem and the expression for the potential energy of the wire are physical statements of the physics of the problem and have no direct correlation with approximation theory.

Consider the original differential equation for the deflection of the tight wire,

$$T\frac{d^2y}{dx^2} + w(x) = 0 \tag{1.51}$$

Because this equation must be satisfied at all points (i.e., must be zero at all points),

$$B(x)\left[T\frac{d^2y}{dx^2} + w(x)\right] = 0 \tag{1.52}$$

must also be satisfied (i.e., equal to zero) for any function $B(x)$. If so, then its integral must likewise equal zero,

$$\int_0^L B(x)\left[T\frac{d^2y}{dx^2} + w(x)\right] dx = 0 \tag{1.53}$$

Interpret this equation as a test that the trial function, $y(x)$, must pass for all permissible test functions, $B(x)$, if it is the exact solution.

We now assume that the test functions, $B(x)$, are sufficiently well behaved that the integrand remains integrable. Thus, we integrate by parts to obtain

$$\int_0^L \left[\frac{d}{dx}\left(BT\frac{dy}{dx} \right) - \frac{dB}{dx}T\frac{dy}{dx} + Bw \right] dx = 0 \tag{1.54}$$

$$BT\left.\frac{dy}{dx}\right]_0^L - \int_0^L \left(\frac{dB}{dx}T\frac{dy}{dx} - Bw \right) dx = 0 \tag{1.55}$$

The integrated first term, in later problems, will allow us to specify boundary conditions other than what we have. For now, however, we assume the original boundary conditions of our tight wire problem, i.e., $y(x)$ known at both $x = 0$ and $x = L$, and let Eq. 1.55 apply only to functions that meet these conditions. This being the case, there is no need to test $y(x)$ at these points, and we restrict our test functions to those that are zero at $x = 0$ and $x = L$. Thus, for our specific problem,

$$\int_0^L \left(\frac{dB}{dx}T\frac{dy}{dx} - Bw \right) dx = 0.0 \tag{1.56}$$

Note that Eq. 1.56 applies to the $y(x)$ that satisfies our boundary conditions at $x = 0$ and $x = L$, whether $y(x)$ is specified as zero or some other value at these points. On the other hand, the test functions $B(x)$ must be zero at these points; otherwise we would have to include the first term in Eq. 1.55, which requires the value of dy/dx. This we ordinarily do not know when y is known. More on this point in the next chapter.

We now have two ways of defining our problem:

1. The deflection of the tight wire, $y(x)$, must satisfy the specified boundary conditions at $x = 0$ and $x = L$, and

$$T\frac{d^2y}{dx^2} + w(x) = 0$$

at every point $0 < x < L$.

2. The deflection of the tight wire, $y(x)$, must satisfy the specified boundary conditions at $x = 0$ and $x = L$, and

$$\int_0^L \left(\frac{dB}{dx}T\frac{dy}{dx} - Bw \right) dx = 0.0$$

for all test functions $B(x)$ for which $B(0) = B(L) = 0$.

The first formulation places stronger demands on the smoothness of the function than does the second formulation; hence, the designations *strong form* and *weak form* are used for the first and second forms, respectively. It will be this second formulation, i.e., the weak formulation, that will be the starting point of all our finite element formulations.

In order to see the equivalence of the above weak form to the potential energy functional used for the Ritz method, we must go back to that formulation, but this time differentiate and then integrate our equation.

$$
\begin{aligned}
\frac{\partial E}{\partial A_1} &= \frac{\partial}{\partial A_i} \int_0^L \left[\frac{1}{2} T \left(\frac{dy}{dx} \right)^2 + wy \right] dx \\
&= \int_0^L \left[T \frac{\partial}{\partial A_i} \left(\frac{dy}{dx} \right) \left(\frac{dy}{dx} \right) + w \left(\frac{\partial y}{\partial A_i} \right) \right] dx \qquad (1.57) \\
&= \int_0^L \left[\frac{d}{dx} \left(\frac{\partial y}{\partial A_i} \right) T \left(\frac{dy}{dx} \right) + w \left(\frac{\partial y}{\partial A_i} \right) \right] dx = 0
\end{aligned}
$$

Comparison of this final form of the Ritz method with Eq. 1.56 shows that they are identical if we use as our test functions

$$
B = \frac{\partial y}{\partial A_i}
$$

Likewise, comparison of Galerkin's formulation (Eqs. 1.23) with Eq. 1.53 shows that they are identical if we, again, use as our test functions

$$
B = \frac{\partial y}{\partial A_i}
$$

Therefore, both Galerkin's method and the Ritz method simply represent use of the weak form of our problem to evaluate the undetermined parameters of an approximate solution.

1.6 SOME CONCLUDING REMARKS

We have assumed that our approximating functions are complete—that as we add more terms, the series can be made as close to the exact solution as we desire. We accepted that the polynomial and trigonometric series are complete. By the very nature of our finite element approximation, it is complete. However, even if we are confident that our approximating function can be made close to the exact solution, how do we know that the methods we have used to determine the parametric value actually force the approximation to be close to the exact solution? In general, this is a difficult question to answer and is best left to texts on that subject. However, it is worthwhile to develop an intuitive feel for why these methods work. A basic characteristic of all four of the methods is that they create a set of linear algebraic equations for the undetermined parameters. This will be true whenever these methods are used to approximate solutions of linear differential equations. Thus, in what follows, keep in mind that each method can produce only one set of values. Let us now review the four methods we introduced in this chapter.

Collocation. Of the methods presented, collocation is perhaps the most believable. As we increase the number of terms in our approximating function, we are able to increase the number of points at which the approximation exactly satisfies the differential equation. Hence, in the limit, among all the functions we are considering, we can both find a function that can duplicate the exact solution and find a function that satisfies the differential equation everywhere. Because there can be only one set of parameters for the function that satisfies the differential equation everywhere, and the exact solution is a function that does the same, the approximating function should approach the exact solution.

Least Squares. Here we seek to find the parametric values of our approximating function that minimize

$$L = \int_0^L R^2 \, dx$$

Because our approximating function can be forced, in the limit, to approach any function, and because the exact function minimizes L, we conclude that by forcing the parameters to minimize L, the approximation must be forced to approach the exact solution.

Galerkin's Method. This is perhaps the most obscure of the methods we studied. What it requires is that our set of parametric values force the following equations to be satisfied:

$$\int_0^L P_1(x)R(x) \, dx = 0$$

$$\int_0^L P_2(x)R(x) \, dx = 0$$

$$\vdots$$

$$\int_0^L P_n(x)R(x) \, dx = 0$$

Here $R(x)$ is the residual and $P_i(x)$ are the independent terms in the approximating function

$$y(x) = \sum_{i=1}^{n} A_i P_i(x)$$

Because each equation is equal to zero, we can multiply each by an arbitrary constant and then sum them; thus

$$\int_0^L B_1 P_1(x)R(x) \, dx = 0$$

$$\int_0^L B_2 P_2(x)R(x) \, dx = 0$$

$$\vdots$$

$$\int_0^L B_n P_n(x)R(x) \, dx = 0$$

and

$$\sum_{i=1}^{n} \int_0^L B_i P_i(x)R(x) \, dx = 0$$

$$\int_0^L \sum_{i=1}^{n} B_i P_i(x)R(x) \, dx = 0$$

$$\int_0^L B(x)R \, dx = 0$$

where

$$B(x) = \sum_{i=1}^{n} B_i P_i(x)$$

Because the parameters are arbitrary and because the series is complete, we can make $B(x)$ approach any function we desire, including $R(x)$. This requires that, in the limit,

$$\int_0^L R^2 \, dx \rightarrow 0$$

Therefore, in the limit $R \rightarrow 0$; thus, our approximating function must approach the exact solution.

The Ritz Method. This is similar to the least squares method in that we force a functional to have a minimum value. In our case the functional was

$$E = \int_0^L \left[\frac{1}{2} T \left(\frac{dy}{dx} \right)^2 + wy \right]$$

Because our approximating function can be forced, in the limit, to approach any function, and because the exact function minimizes E, then forcing the parameters to minimize E should likewise force the approximation to approach the exact solution.

The preceding arguments do not constitute rigorous proofs for convergence, but they do point to how such proofs are created and, therefore, provide insight as to how these methods work. Such insight will be helpful in your understanding the material to follow as well as in your later use of the finite element method in whatever capacity you choose.

EXERCISES

Study Problems

S1. Consider another application associated with the example problem, such as heat conduction, porous flow, or electrostatics, and identify the physical significance of T and W in the governing equation.

S2. For each of the methods, use only one degree of freedom in the polynomial series, e.g.,

$$y(x) = (x)(x - L)A_0$$

and determine the "best" value for A_0 that each method gives. Use the loading shown in Fig. 1.2.

S3. For each of the methods, use the approximating function

$$y(x) = A_0 \sin\left(\frac{\pi x}{L}\right)$$

and determine the "best" value for A_0 that each method gives. Use the loading shown in Fig. 1.2.

S4. For the loading shown in Fig. 1.2, determine the potential energy of the tight wire problem corresponding to the exact solution. Give your answer in terms of W^2L^3/T.

S5. For the loading shown in Fig. 1.2, determine the potential energy that the two-parameter approximate solutions give to the tight wire problem. Give your answer in terms of W^2L^3/T.

S6. Consider the loading on the wire constant over its entire length and derive the exact solution. Show that all four methods give the exact solution with the approximating function

$$y(x) = (x)(x - L)A_0$$

S7. Fully explain why the finite element series is complete. What qualifying statement would you have to make concerning how the series is made?

Numerical Experiments and Code Development

N1. For the tight wire problem with a constant uniform load equal to W, and $T = 100WL$, calculate the potential energy as E/WL^2 using

$$y(x) = A_0 \sin\left(\frac{\pi x}{L}\right)$$

and plot it as a function of A_0/L. On the same plot, indicate the potential energy that the exact solution gives.

N2. For the tight wire problem with a constant uniform load equal to W, calculate

$$J = \int_0^L R^2 \, dX$$

using

$$y(x) = A_0 \sin\left(\frac{\pi x}{L}\right)$$

and plot J/W^2L as a function of A_0/L. Let $T = 100WL$. What is the value of J that the exact solution gives?

N3. For the tight wire problem with a constant load equal to W, use the approximating function

$$y = \sum_{n=1}^{N} A_n \sin\left(\frac{n\pi x}{L}\right)$$

and write a computer program to calculate the values of A_n by collocation for any N. Plot $(y_{exact} - y_{approx})$ as a function of x/L for $n = 1, 2, 3,$ and 4.

CHAPTER 2

CALCULUS OF VARIATIONS

In this chapter we set forth some of the basic concepts of the calculus of variations that we will use in developing the finite element method. In the previous chapter we defined the potential energy associated with the tight wire in terms of its deflection $y(x)$. The potential energy, therefore, is a scalar function of the function $y(x)$ and is referred to as a functional. If the deflection of the wire is changed a small amount, the potential energy will likewise change. The relationship between a change in a function $y(x)$ and the corresponding change in its dependent functional is the subject matter of the calculus of variations. It has a marked similarity to differential calculus, where one wishes to determine the relationship between a change in a scalar x and the differential of the function $y(x)$, i.e., $dy = (dy/dx)\,dx$.

2.1 THREE FUNDAMENTAL RELATIONSHIPS

We begin by defining the notation used for a change (or variation) in y as $\delta y = \epsilon \eta(x)$. Here η is a function of x, and ϵ is a scalar that we use to control the magnitude of the variation. Thus,

$$y + \delta y = y + \epsilon \eta(x) \tag{2.1}$$

These functions can be visualized as shown in Fig. 2.1.

In the following, we let $y(x)$ represent the (unknown) solution to a specific problem and $\epsilon \eta(x)$ an arbitrary variation of this solution. Usually we will study only small variations of $y(x)$, such that

$$\epsilon^2 \ll |\epsilon| \tag{2.2}$$

We will follow the practice that the symbol δy means that the variation is small in the sense of Eq. 2.2.

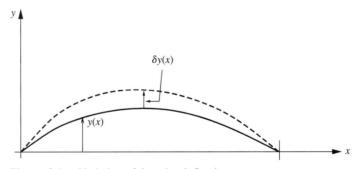

Figure 2.1. Variation of the wire deflection.

Now let us look at some important relationships. We first consider the derivative of $y(x)$ and its variation due to a variation of $y(x)$. By definition, the variation of the derivative is

$$\delta \frac{dy}{dx} = \frac{d(y + \delta y)}{dx} - \frac{dy}{dx} \tag{2.3}$$

which, when expanded, gives us

$$\delta \frac{dy}{dx} = \frac{dy}{dx} + \frac{d(\delta y)}{dx} - \frac{dy}{dx} \tag{2.4}$$

Therefore,

$$\delta \frac{dy}{dx} = \frac{d}{dx} \delta y \tag{2.5}$$

That is, the variation of the derivative is equal to the derivative of the variation. We will use the interchangeability of these operations a great deal in developing our finite element equations.

Next we consider the integral of $y(x)$ and its variation due to a variation of $y(x)$. By definition, we have

$$\delta \left[\int y(x)\,dx \right] = \left[\int [y(x) + \delta y(x)]\,dx \right] - \left[\int [y(x)]\,dx \right] \tag{2.6}$$

Upon expanding the second integral, we obtain

$$\delta \left[\int y\,dx \right] = \left[\int y\,dx \right] + \left[\int \delta y\,dx \right] - \left[\int y\,dx \right] \tag{2.7}$$

which is simply

$$\delta \int y\,dx = \int \delta y\,dx \tag{2.8}$$

That is, the variation of the integral is equal to the integral of the variation. Thus, the operations of integration and variation are interchangeable.

Now we consider a function $F(y)$, where $y = y(x)$, and pose the question, how does the function F change when $y(x)$ is changed from one function to another? That is, suppose the function

$$y(x) = Y(x) \tag{2.9}$$

is changed to the function

$$y(x) = Y(x) + \epsilon \eta(x) \tag{2.10}$$

Then the change in $F(y(x))$ would be

$$\delta F(y) = F\left[Y(x) + \epsilon \eta(x)\right] - F\left[Y(x)\right] \tag{2.11}$$

The first term on the right-hand side can be expanded in a Taylor's series about $\epsilon = 0$ to obtain

$$
\begin{aligned}
F(y) &= F\left[Y(x) + \epsilon\eta(x)\right] \\
&= F\big|_{y=Y} + \frac{dF}{d\epsilon}\bigg|_{y=Y}\epsilon + \frac{1}{2}\frac{d^2F}{d\epsilon^2}\bigg|_{y=Y}\epsilon^2 + O(\epsilon^3)
\end{aligned}
\tag{2.12}
$$

For small ϵ, we may neglect the higher-order terms in ϵ and write

$$
\delta F = F(Y) + \frac{dF(Y)}{d\epsilon}\epsilon - F(Y)
\tag{2.13}
$$

or simply

$$
\delta F = \frac{dF}{d\epsilon}\epsilon
\tag{2.14}
$$

The derivative can be written as

$$
\frac{dF}{d\epsilon} = \frac{dF}{dy}\frac{dy}{d\epsilon} = \frac{dF}{dy}\eta
\tag{2.15}
$$

where we have used

$$
y = Y + \epsilon\eta
\tag{2.16}
$$

Hence,

$$
\delta F = \left(\frac{dF}{dy}\right)\eta\epsilon
\tag{2.17}
$$

Because $\eta\epsilon$ represents the small change of the argument y (see Eq. 2.1) of the function $F(y)$, we can write

$$
\delta F = \left(\frac{dF}{dy}\right)\delta y
\tag{2.18}
$$

We have thus obtained an expression for the infinitesimal variation of the function $F(y)$ due to an infinitesimal change in the function $y(x)$, which is completely analogous to the expression for a differential that we have from differential calculus.

2.2 APPLICATION TO THE TIGHT WIRE PROBLEM

We now apply the result of the previous section to our tight wire problem. In Chapter 1, two formulations were given for the solution of this problem: the strong form, which stated that the deflection must satisfy

$$T\frac{d^2y}{dx^2} + w(x) = 0.0 \tag{2.19}$$

at every point, and the weak form, which stated that the deflection must give a stationary value to the functional

$$E(y) = \int_0^L \left[\frac{1}{2}T\left(\frac{dy}{dx}\right)^2 - wy \right] dx \tag{2.20}$$

We now demonstrate the equivalence of these two formulations by showing

1. If $y(x)$ satisfies Eq. 2.19, then when it is substituted into Eq. 2.20 and given a small variation $\delta y(x)$, there will be no variation of $E(y)$, i.e., $\delta E = 0$.
2. If $y(x)$ gives a stationary value to $E(y)$, i.e., $\delta E = 0$ for any δy, then $y(x)$ satisfies Eq. 2.19.

We begin by taking the variation of E with respect to y.

$$\delta E(y) = \delta \int_0^L \left[\frac{1}{2}T\left(\frac{dy}{dx}\right)^2 - wy \right] dx \tag{2.21}$$

Interchanging the operations of variation and integration, we obtain

$$\delta E(y) = \int_0^L \delta \left[\frac{1}{2}T\left(\frac{dy}{dx}\right)^2 - wy \right] dx \tag{2.22}$$

Application of the standard formula of differential calculus gives us

$$\delta E(y) = \int_0^L \left[T\left(\frac{dy}{dx}\right)\delta\left(\frac{dy}{dx}\right) - w\,\delta y \right] dx \tag{2.23}$$

We now interchange the operations of variation and differentiation to obtain

$$\delta E(y) = \int_0^L \left[T\left(\frac{dy}{dx}\right)\frac{d}{dx}(\delta y) - w\,\delta y \right] dx \tag{2.24}$$

To arrive at this point we used all three relationships developed in the previous section. From here on, however, we need only to follow the rules of differential and integral calculus. We next integrate Eq. 2.24 by parts; that is, we use

$$\frac{d(AB)}{dx} = \frac{dA}{dx}B + \frac{dB}{dx}A \tag{2.25}$$

to obtain

$$\frac{dy}{dx}\frac{d}{dx}(\delta y) = \frac{d}{dx}\left(\frac{dy}{dx}\delta y\right) - \frac{d^2y}{dx^2}\delta y \tag{2.26}$$

Thus,

$$\delta E = \int_0^L \left[T\frac{d}{dx}\left(\frac{dy}{dx}\delta y\right) - T\frac{d^2y}{dx^2}\delta y - w\,\delta y \right] dx \tag{2.27}$$

The first term can be integrated to give

$$\delta E = \left. T\frac{dy}{dx}\delta y \right|_0^L - \int_0^L \left[T\frac{d^2y}{dx^2}\delta y + w\,\delta y \right] dx \tag{2.28}$$

It is important to note two limitations that we must now place on the variation δy. First, in order to perform the integration indicated in the previous step, the integrand,

$$T\frac{d}{dx}\left(\frac{dy}{dx}\delta y\right)$$

must be integrable. Thus, δy must be a continuous function: otherwise its derivative would not exist at pints of discontinuity. Second, we must require that δy at the endpoints be zero; that is,

$$\delta y(0) = \delta y(L) = 0.0 \tag{2.29}$$

Otherwise the variation would create a function that did not satisfy the boundary conditions. Substitution of these limits of integration makes the first term in Eq. 2.28 equal to zero, and we are left with

$$\delta E = -\int_0^L \left[T\frac{d^2y}{dx^2} + w \right]\delta y\,dx \tag{2.30}$$

Note that if Eq. 2.19 is satisfied, then δE is equal to zero for any δy. Therefore, Eq. 2.19 is a sufficient condition to ensure that $y(x)$ gives a stationary value to E. On the other hand, if δE as given by Eq. 2.30 is zero for any δy (which satisfies the boundary conditions), then the term in braces must be identically zero. Hence, if we find a function $y(x)$ that gives a stationary value to E, then we have found the $y(x)$ that satisfies Eq. 2.19. An equation such as Eq. 2.19, when obtained in this manner, is referred to as the *Euler equation* of the corresponding functional.

A Second Approach. Although the above procedure for obtaining the equation for $y(x)$ that produces a stationary value of E is direct and easy to use, it is instructive, and often useful, to use the following approach.

Consider the potential energy written in terms of $y(x)$ (the function that gives it a stationary value) and its variation $\epsilon \eta$,[1]

$$E(y + \epsilon\eta) = \int_0^L \left[\frac{1}{2}T\left[\frac{d}{dx}(y + \epsilon\eta)\right]^2 - w(y + \epsilon\eta) \right] dx \tag{2.31}$$

Because E has a stationary value at $\epsilon = 0$, we know

$$\left.\frac{\partial E}{\partial \epsilon}\right|_{\epsilon=0} = 0 \tag{2.32}$$

for any $\eta(x)$. Expansion of Eq. 2.31 gives the following quadratic in ϵ.

$$E(y + \epsilon\eta) = \left[\int_0^L \left[\frac{1}{2}T\left(\frac{dy}{dx}\right)^2 - wy \right] dx \right]$$

$$+ \left[\int_0^L \left[T\frac{dy}{dx}\frac{d\eta}{dx} - w\eta \right] dx \right] \epsilon$$

$$+ \left[\int_0^L \left[\frac{1}{2}T\left(\frac{d\eta}{dx}\right)^2 \right] dx \right] \epsilon^2 \tag{2.33}$$

The derivative of E with respect to ϵ is

$$\frac{\partial E}{\partial \epsilon} = \left[\int_0^L \left[T\frac{dy}{dx}\frac{d\eta}{dx} - w\eta \right] dx \right]$$

$$+ \left[\int_0^L \left[\frac{1}{2}T\left(\frac{d\eta}{dx}\right)^2 \right] dx \right] \epsilon \tag{2.34}$$

If this is to be zero at $\epsilon = 0$, then

$$\int_0^L \left[T\frac{dy}{dx}\frac{d\eta}{dx} - w\eta \right] dx = 0 \tag{2.35}$$

You should now note the similarity between Eq. 2.35 and Eq. 2.24. These two equations impose identical requirements since the integrals must be zero for either an arbitrary δy or an arbitrary η. We therefore arrive at the same governing equation for $y(x)$ as we had before.

This second approach, however, gives us one bit of information that the first approach does not. Consider the expression for $E(y + \epsilon\eta)$ as given by Eq. 2.33. The second integral on the right-hand side we now know equals zero at the stationary value of E. Hence, $E(y + \epsilon\eta)$ is the sum of the E in the equilibrium position (the first integral) and an integral that is always positive. The stationary value of E is, therefore, a minimum value for the potential energy of the system.

---

[1] Remember, $\eta(0) = \eta(L) = 0$, so that the variation of y produces a new function that still satisfies the specified boundary conditions.

2.3 CORRESPONDING FUNCTIONALS

For the tight wire problem there existed a potential energy that produced a minimum value for the exact solution. Thus, we knew that for the differential equation

$$T\frac{d^2y}{dx^2} + w(x) = 0.0$$

there existed the corresponding functional

$$E(y) = \int_0^L \left[\frac{1}{2}T\left(\frac{dy}{dx}\right)^2 - wy\right]dx$$

In the previous section we demonstrated that for a given functional, the corresponding differential equation (the Euler equation) can easily be obtained. We now want to consider the following three questions:

1. Given a differential equation, is there always such a corresponding functional?
2. If there is a corresponding functional, can it be obtained from the differential equation?
3. If there is not a corresponding functional, can the finite element method be used to obtain approximate solutions to the governing equation?

Unfortunately, the answer to the first question is no; fortunately, however, the answer to the second and third questions is yes. To arrive at these answers we use the weak formulation as described in Chapter 1. Let us again go through the steps necessary to place the tight wire problem in its weak form, but now use the notation associated with the calculus of variations; that is, we will interpret the arbitrary weighting function as an arbitrary variation in y.

Given the function $y(x)$ that satisfies the above differential equation, $0 \le x \le L$, we know

$$\int_0^L \delta y\left(T\frac{d^2y}{dx^2} + w(x)\right)dx = 0 \tag{2.36}$$

must be satisfied for any variation of the solution δy. Thus,

$$\int_0^L \delta y\left(T\frac{d^2y}{dx^2} + w\right)dx = \int_0^L \left[\frac{d}{dx}\left(\delta yT\frac{dy}{dx}\right) - \frac{d\delta y}{dx}T\frac{dy}{dx} + w\,\delta y\right]dx \tag{2.37}$$

$$= \left[\delta yT\frac{dy}{dx}\right]_0^L - \int_0^L \left[\frac{d\delta y}{dx}T\frac{dy}{dx} - w\,\delta y\right]dx \tag{2.38}$$

$$= -\int_0^L \left[\frac{d\delta y}{dx}T\frac{dy}{dx} - w\,\delta y\right]dx \tag{2.39}$$

$$= - \int_0^L \left[\delta \frac{dy}{dx} T \frac{dy}{dx} - w \, \delta y \right] dx \qquad (2.40)$$

$$= - \int_0^L \left[\delta \left(\frac{1}{2} T \left(\frac{dy}{dx} \right)^2 \right) - \delta(wy) \right] dx \qquad (2.41)$$

$$= - \int_0^L \delta \left[\frac{1}{2} T \left(\frac{dy}{dx} \right)^2 - wy \right] dx \qquad (2.42)$$

$$= - \delta \int_0^L \left[\frac{1}{2} T \left(\frac{dy}{dx} \right)^2 - wy \right] dx \qquad (2.43)$$

$$= - \delta E = 0 \qquad (2.44)$$

We have arrived at the corresponding variational principle of our problem, starting with the original differential equation. Similar steps can be taken to determine the functional corresponding to other differential equations—provided, of course, that a functional does exist.

Where, then, in the above steps was the crucial point that led to there being a functional? It occurred between Eqs. 2.40 and 2.41, where we rewrote each term in the integrand as a total variation of a single function. This allowed us to write the total integrand as a variation of a single function, and finally the entire expression as a variation of an integral.

There are, however, differential equations for which corresponding functionals such as that above do not exist. Such would be the case if, in our original linear differential equation, the first derivative of y had been present. In that case, the integrand in Eq. 2.40 would have included the term

$$\left(\frac{dy}{dx} \right) \delta y \quad \text{or simply} \quad y' \, \delta y$$

The fact that there is no function, say $J(y, y')$, whose total variation equals the above is easily shown. The total variation of J would be

$$\delta J = \frac{\partial J}{\partial y} \delta y + \frac{\partial J}{\partial y'} \delta y' \qquad (2.45)$$

But this would have to equal

$$\delta J = y' \, \delta y \qquad (2.46)$$

Comparison of the two forms for δJ shows that the partial of J with respect to y' must be zero; hence, J can not be a function of y'. This, however, contradicts Eq. 2.46; thus, J does not exist.

In comparison, consider that if our original equation had contained the function itself, e.g.,

$$\frac{d^2y}{dx^2} + y = 0$$

then $y\,\delta y$ would have appeared in the integrand. However, in this case, we could have written

$$y\,\delta y = \delta\left(\tfrac{1}{2}y^2\right)$$

to have created the necessary total variation.

We have thus demonstrated that when there is a corresponding functional, it can be derived from the differential equation; however, there may not always be a corresponding functional. Thus, we turn to our final question: If there is not a corresponding functional, can we still obtain an approximate solution by the finite element method? The answer, as stated above, is yes. We do not need a functional in order to use our finite element approximations. In fact, when there is such a functional, the first thing we do is take its variation to arrive at an equation similar to Eq. 2.40. It is this form into which we substitute our finite element approximations. Thus, the steps leading to Eqs. 2.41–2.44 are unnecessary for the development of our finite element equations. In the remainder of this book we will follow steps similar to those leading to Eq. 2.40 to formulate the governing equation used for the finite element approximation. It is this formulation that we will refer to as the weak form of the problem.

One final comment (question): Why, if we do not need a functional in order to develop our finite element equations, did we say it was unfortunate that the answer to our first question was no? There are two reasons. The first reason is that even if the functional is not directly used in the development of our equations, the fact that it exists gives us a better understanding of how our method converges to the solution (see comments at end of Chapter 1). The second reason is that when a functional does exist, the resulting finite element equations are described in terms of symmetric algebraic equations, whereas when a functional does not exist, the resulting equations are nonsymmetric.

EXERCISES

Study Problems

S1. Given

$$y = \sin(x) \quad \text{and} \quad \delta y = -0.5 + 0.2x^2$$

(a) Compare

$$\frac{d}{dx}(y + \delta y) - \frac{d}{dx}(y) \quad \text{with} \quad \frac{d}{dx}(\delta y)$$

(b) Compare

$$\int_0^3 (y + \delta y)dx - \int_0^3 (y)dx \quad \text{with} \quad \int_0^3 (\delta y)dx$$

S2. Given

$$F = 2y^2$$

let

$$y(x) = Y(x) = 6\sin(x)$$

Give a variation to $Y(x)$ equal to

$$\delta Y(x) = -\epsilon x^2$$

Determine by expanding all terms

$$\delta F = F(Y + \delta Y) - F(Y)$$

Write δF after all terms containing ϵ^2 and higher have been neglected. Now calculate

$$\frac{dF}{dy} \quad \text{at} \quad y = Y$$

Compare

$$\delta F = \frac{dF}{dy}\delta Y$$

with your previous equation for δF.

Numerical Experiments and Code Development

N1. Given

$$y = x$$

$$F(y) = y\cos(y)$$

$$\delta y = \epsilon\eta = \epsilon\left[5.0 - (x - 1.5)^2\right]$$

write a program to plot $F(y + \delta Y) - F(y)$ and $(dF/dy)\delta y$ as functions of x in the range $0 \le x \le \pi$ for $\epsilon = 1.0, 0.5, 0.1$, and 0.01. Discuss your solutions in terms of the approximations given in the text.

Answer:

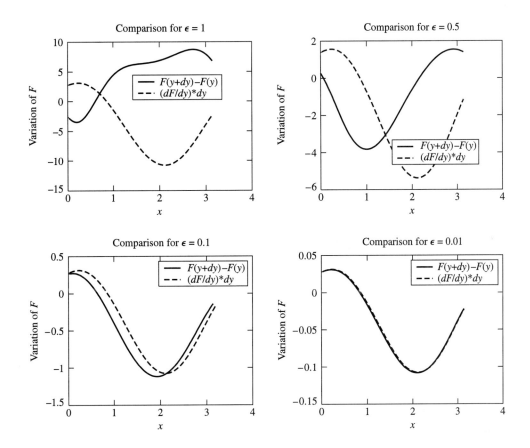

CHAPTER 3

A FINITE ELEMENT PROGRAM

We now apply the principles of the previous chapters to develop a finite element program for the tight wire problem. All steps, from the development of the finite element equations to the writing of a complete finite element program, will be presented, setting the foundations for the remainder of this text.

3.1 FOUR SIMPLIFYING TECHNIQUES

In developing the finite element equations, the variational formulation of the problem will be used along with the following four simplifying techniques:

1. Rather than substitute our approximation into the potential energy functional, we will substitute it into the variation of the functional:

$$\delta E = \int_0^L \left[T \frac{dy}{dx} \delta \frac{dy}{dx} - w \delta y \right] dx \tag{3.1}$$

This corresponds to starting with the weak formulation as described in Chapter 1. We will use this approach in other chapters as well.

2. All equations for a general element will be developed using an approximating function given in terms of a local coordinate system with origin at the left end of the element. Hence,

$$y(x) = Y_a + (Y_b - Y_a)\frac{x}{L_e} \tag{3.2}$$

where the terms are defined in Fig. 3.1. Notice that this is simply a shift in origin of the coordinate system and does not affect the shape functions or their derivatives. It is done simply to facilitate the integration.

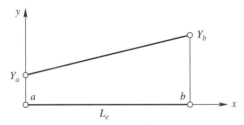

Figure 3.1. A single-element approximation.

3. We will express all our equations using matrix notation. Hence, the above approximation for y is written

$$y = \left\lfloor \left(1 - \frac{x}{L_e}\right) \quad \left(\frac{x}{L_e}\right) \right\rfloor \left\{ \begin{matrix} Y_a \\ Y_b \end{matrix} \right\} \tag{3.3}$$

4. We will include known boundary values as if they were unknowns in the formulation and make correction for this after the final matrix equation is completed.

3.2 SHAPE FUNCTIONS

We now introduce an important terminology associated with the finite element method. In Eq. 3.3, the two functions in the row matrix are referred to as the *shape functions* for the element. They are often abbreviated using the notation N:

$$y(x) = \lfloor N_a \quad N_b \rfloor \begin{Bmatrix} Y_a \\ Y_b \end{Bmatrix}$$

(3.4)

You should note that each shape function is associated with a corresponding node. Note also that N_i, the function associated with node i, has a value of unity at node i and is zero at the other node, as shown in Fig. 3.2.

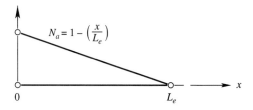

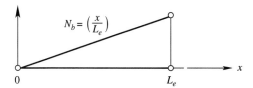

Figure 3.2.   Shape functions.

Before proceeding, we note that in addition to the relationship expressed by Eq. 3.3, we have

$$\frac{dy}{dx} = \left\lfloor \left(-\frac{1}{L_e}\right) \quad \left(\frac{1}{L_e}\right) \right\rfloor \begin{Bmatrix} Y_a \\ Y_b \end{Bmatrix}$$

(3.5)

$$\delta y = \left\lfloor \left(1 - \frac{x}{L_e}\right) \quad \left(\frac{x}{L_e}\right) \right\rfloor \begin{Bmatrix} \delta Y_a \\ \delta Y_b \end{Bmatrix}$$

(3.6)

$$\frac{d}{dx}(\delta y) = \left\lfloor \left(-\frac{1}{L_e}\right) \quad \left(\frac{1}{L_e}\right) \right\rfloor \begin{Bmatrix} \delta Y_a \\ \delta Y_b \end{Bmatrix}$$

(3.7)

It is important to note here that the derivatives indicated are those shown in the original equations and should be understood as being derivatives with respect to the global coordinates. However, the local coordinate system we now associate with x has the same length scale as the global coordinate system; i.e., other than a shift in origin, it is the same as the global coordinate system. Hence, the derivative of the shape function with respect to the global coordinate is the same as it is with respect to the local coordinate, and we need not make a distinction between the two derivatives.

   The following shorthand notation for the preceding equations will be used:

$$
\boxed{
\begin{aligned}
y(x) &= \lfloor N \rfloor \{Y\} \\[4pt]
\frac{dy}{dx} &= \lfloor N_x \rfloor \{Y\} \\[4pt]
\delta y &= \lfloor N \rfloor \{\delta Y\} \\[4pt]
\frac{d}{dx}(\delta y) &= \delta \frac{dy}{dx} = \lfloor N_x \rfloor \{\delta Y\}
\end{aligned}
}
\qquad (3.8)
$$

3.3 ELEMENT STIFFNESS

The next step in our development is to substitute the preceding finite element approximation into the expression for the variation in the potential energy, Eq. 3.1. The integration is performed element by element, i.e.,

$$
\delta E = \int_0^L \left[T \frac{dy}{dx} \delta \frac{dy}{dx} - w \, \delta y \right] dx = \sum_1^N \int_0^{L_e} \left[T \frac{dy}{dx} \delta \frac{dy}{dx} - w \, \delta y \right] dx
\qquad (3.9)
$$

where N is the number of elements and the limits of integration, 0 to L_e, imply that x is measured from the left node of each element in accordance with our local coordinate system.

This procedure, integrating our variational equation element by element, is the basic building block of the finite element method. Computer codes are written to do just this, looping over the entire number of elements until the complete variation has been integrated. Because local coordinates are used for each element, the algorithm for this integration is exactly the same for each element. We therefore consider the integral for a single generic element and label its contribution to the total integration of δE as δE_e. Thus,

$$
\delta E_e = \int_0^{L_e} \left[T \frac{dy}{dx} \delta \frac{dy}{dx} - w \, \delta y \right] dx
$$

$$
= \int_0^{L_e} \lfloor \delta Y_a \quad \delta Y_b \rfloor
\left\{
\begin{array}{c}
-\dfrac{1}{L_e} \\[6pt]
\dfrac{1}{L_e}
\end{array}
\right\}
T \left\lfloor -\dfrac{1}{L_e} \quad \dfrac{1}{L_e} \right\rfloor
\left\{
\begin{array}{c}
Y_a \\[4pt]
Y_b
\end{array}
\right\} dx
$$

$$
- \int_0^{L_e} \lfloor \delta Y_a \quad \delta Y_b \rfloor
\left\{
\begin{array}{c}
1 - \dfrac{x}{L_e} \\[6pt]
\dfrac{x}{L_e}
\end{array}
\right\} w(x) \, dx
\qquad (3.10)
$$

Because Y_a, Y_b, δY_a, and δY_b are nodal point values and not functions of x, they may be moved outside the integral and not considered part of the integrand. Although some of the other terms are also constant, in later applications they will not be and should be considered part of the integrand. Integration gives us

$$\delta E_e = \lfloor \delta Y_a \quad \delta Y_b \rfloor \begin{bmatrix} +\dfrac{T}{L_e} & -\dfrac{T}{L_e} \\ -\dfrac{T}{L_e} & +\dfrac{T}{L_e} \end{bmatrix} \begin{Bmatrix} Y_a \\ Y_b \end{Bmatrix} - \lfloor \delta Y_a \quad \delta Y_b \rfloor \begin{Bmatrix} F_a \\ F_b \end{Bmatrix} \tag{3.11}$$

where

$$F_a = \int_0^{L_e} \left(1 - \frac{x}{L_e}\right) w(x)\, dx$$

$$\tag{3.12}$$

$$F_b = \int_0^{L_e} \left(\frac{x}{L_e}\right) w(x)\, dx$$

Clearly, these last terms depend on the particular loading, $w(x)$, specified for the problem.

It will often be convenient to write the expression for δE_e in the following abbreviated notation:

$$\delta E_e = \lfloor \delta Y \rfloor_e [K]_e \{Y\}_e - \lfloor \delta Y \rfloor_e \{F\}_e$$

where

$$[K]_e = \begin{bmatrix} +\dfrac{T}{L_e} & -\dfrac{T}{L_e} \\ -\dfrac{T}{L_e} & +\dfrac{T}{L_e} \end{bmatrix}$$

$$\tag{3.13}$$

$$\{F\}_e = \begin{Bmatrix} \displaystyle\int_0^{L_e} \left(1 - \frac{x}{L_e}\right) w(x)\, dx \\ \displaystyle\int_0^{L_e} \left(\frac{x}{L_e}\right) w(x)\, dx \end{Bmatrix}$$

3.4 THE GLOBAL STIFFNESS MATRIX

Equations 3.13 represent the contribution to δE from a generic element e. We now consider how these contributions are added together to give the total variation δE. The form of the equation for the total δE will be exactly the same as that for a single element, that is,

$$\delta E = \lfloor \delta Y \rfloor [K]\{Y\} - \lfloor \delta Y \rfloor \{F\} \tag{3.14}$$

but now $[K]$ will be an $N \times N$–dimensional matrix, and $\{Y\}$, $\{\delta Y\}$, and $\{F\}$ will be N-dimensional vectors, where N equals the total number of nodes.

To show how this addition is performed, we consider that our wire is divided into three elements of lengths L_1, L_2, and L_3. The total variation will be

$$\delta E = \delta E_1 + \delta E_2 + \delta E_3 \tag{3.15}$$

Now note that each element contribution can be written in terms of the total number of nodal points. For example, the scalar equation for δE_2,

$$\delta E_2 = \lfloor \delta Y_2 \quad \delta Y_3 \rfloor \begin{bmatrix} +\dfrac{T}{L_2} & -\dfrac{T}{L_2} \\[2mm] -\dfrac{T}{L_2} & +\dfrac{T}{L_2} \end{bmatrix} \left\{ \begin{matrix} Y_2 \\ Y_3 \end{matrix} \right\} + \lfloor \delta Y_2 \quad \delta Y_3 \rfloor \left\{ \begin{matrix} F_{a2} \\ F_{b2} \end{matrix} \right\} \tag{3.16}$$

is equivalent to

$$\delta E_2 = \lfloor \delta Y_1 \quad \delta Y_2 \quad \delta Y_3 \quad \delta Y_4 \rfloor \begin{bmatrix} 0 & 0 & 0 & 0 \\[2mm] 0 & +\dfrac{T}{L_2} & -\dfrac{T}{L_2} & 0 \\[2mm] 0 & -\dfrac{T}{L_2} & +\dfrac{T}{L_2} & 0 \\[2mm] 0 & 0 & 0 & 0 \end{bmatrix} \left\{ \begin{matrix} Y_1 \\ Y_2 \\ Y_3 \\ Y_4 \end{matrix} \right\} + \lfloor \delta Y_1 \quad \delta Y_2 \quad \delta Y_3 \quad \delta Y_4 \rfloor \left\{ \begin{matrix} 0 \\ F_{a2} \\ F_{b2} \\ 0 \end{matrix} \right\} \tag{3.17}$$

where L_2 equals the length of the second element, and F_{a2} and F_{b2} represent the left and right contributions to $\{F\}$ from element 2 as obtained from Eqs. 3.12.

If each element's contribution is written in this manner, their sum is easily obtained by adding their stiffness matrices and each of the right-hand-side vectors. The total, or global, stiffness matrix becomes

$$[K] = \begin{bmatrix} \dfrac{T}{L_1} & -\dfrac{T}{L_1} & 0 & 0 \\[3mm] -\dfrac{T}{L_1} & \dfrac{T}{L_1}+\dfrac{T}{L_2} & -\dfrac{T}{L_2} & 0 \\[3mm] 0 & -\dfrac{T}{L_2} & \dfrac{T}{L_2}+\dfrac{T}{L_3} & -\dfrac{T}{L_3} \\[3mm] 0 & 0 & -\dfrac{T}{L_3} & \dfrac{T}{L_3} \end{bmatrix} \tag{3.18}$$

The total, or global, right-hand-side matrix becomes

$$\{F\} = \left\{ \begin{array}{c} F_{a1} \\ \hline F_{b1} + F_{a2} \\ \hline F_{b2} + F_{a3} \\ \hline F_{b3} \end{array} \right\} \tag{3.19}$$

Our variational principle tells us that the variation, δE, must be zero for any arbitrary variation of the nodal point values, δY, that are consistent with our boundary conditions. For our particular problem, this means that δY_1 and δY_4 would have to be zero in order not to change our approximation to one that did not satisfy the known boundary conditions. However, we will allow these variations to be nonzero and make adjustments later to ensure that our final solution produces the specified boundary values. Thus,

$$\delta E = \lfloor \delta Y \rfloor [K] \{Y\} - \lfloor \delta Y \rfloor \{F\} = 0 \tag{3.20}$$

This appears to be a single equation, but it is in fact much more than that because it must be true for arbitrary δY's. To show fully the consequences of this, we write the equation as

$$\lfloor \delta Y \rfloor \{ [K]\{Y\} - \{F\} \} = 0.0 \tag{3.21}$$

This equation represents the scalar product of two vectors, and it states that this product must be zero for any vector $\lfloor \delta Y \rfloor$. This can be true only if the second vector is identically zero. Hence, we conclude

$$[K]\{Y\} - \{F\} = 0.0 \tag{3.22}$$

more often written as

$$[K]\{Y\} = \{F\} \tag{3.23}$$

Equation 3.23 represents the final matrix equation for the deflected shape. However, it is in error due to the fact that we allowed arbitrary variations of Y at the ends of the wire. Thus, as it now stands, we have not restricted our set of possible solutions to those that satisfy the prescribed boundary conditions. We now show how to take care of this problem.

3.5 BOUNDARY CONDITIONS

As things now stand, the first and last equations associated with the matrix equations are invalid; they imply governing relationships for the end deflections that are not equivalent to the requirements that they be equal to the specified deflections. Hence, in order to proceed, we must remove this contradiction and specify the end values in accordance with the derivation of the potential energy for the tight wire problem. There are three possible methods for doing this.

Method 1. The first method, and perhaps the most obvious, is to simply remove the first and last equations associated with the end values, and move the corresponding columns to the right-hand side. For our three-element example, we would then have

$$
\left[
\begin{array}{c|c}
\dfrac{T}{L_1} + \dfrac{T}{L_2} & -\dfrac{T}{L_2} \\
\hline
-\dfrac{T}{L_2} & \dfrac{T}{L_2} + \dfrac{T}{L_3}
\end{array}
\right]
\left\{
\begin{array}{c}
Y_2 \\ \hline Y_3
\end{array}
\right\}
=
\left\{
\begin{array}{c}
F_2 \\ \hline F_3
\end{array}
\right\}
-
\left[
\begin{array}{c|c}
-\dfrac{T}{L_1} & 0 \\
\hline
0 & -\dfrac{T}{L_3}
\end{array}
\right]
\left\{
\begin{array}{c}
\tilde{Y}_1 \\ \hline \tilde{Y}_4
\end{array}
\right\}
$$

or simply

$$
\left[
\begin{array}{c|c}
\dfrac{T}{L_1} + \dfrac{T}{L_2} & -\dfrac{T}{L_2} \\
\hline
-\dfrac{T}{L_2} & \dfrac{T}{L_2} + \dfrac{T}{L_3}
\end{array}
\right]
\left\{
\begin{array}{c}
Y_2 \\ \hline Y_3
\end{array}
\right\}
=
\left\{
\begin{array}{c}
F_2 + \dfrac{T}{L_1}(\tilde{Y}_1) \\ \hline F_3 + \dfrac{T}{L_3}(\tilde{Y}_4)
\end{array}
\right\}
\tag{3.24}
$$

In our example the known end deflections are zero; however, it is permissible to specify any value desired. Hence, we have indicated these known values as $\tilde{Y}_1$ and $\tilde{Y}_4$. This approach has both an advantage and a disadvantage. Clearly, the advantage is that the size of the matrix equation to be solved has been reduced. Although for this simple example the reduction was minimal, for two- and three-dimensional problems the number of specified boundary values (and hence the number of equations to be eliminated) can be in the hundreds. The disadvantage of this approach occurs when the number of equations (and hence the number of rows and columns of the stiffness matrix) to be eliminated is large and appear at various locations within the stiffness matrix. In such cases, their elimination can involve a difficult rearrangement of the stiffness matrix. This disadvantage, however, can be partially overcome by eliminating them at the time the element matrices in which they appear are placed into the global stiffness matrix. Thus, they never appear in the global matrix to begin with. Some codes use this approach very successfully.

Method 2. The second method is never used but prepares us for the third method, which is often used. This second method is to leave the two known deflections as unknowns in the matrix equation and rewrite the first and last equations to specify their values. If we did this, we would have

$$
\left[
\begin{array}{c|c|c|c}
1 & 0 & 0 & 0 \\
\hline
-\dfrac{T}{L_1} & \dfrac{T}{L_1} + \dfrac{T}{L_2} & -\dfrac{T}{L_2} & 0 \\
\hline
0 & -\dfrac{T}{L_2} & \dfrac{T}{L_2} + \dfrac{T}{L_3} & -\dfrac{T}{L_3} \\
\hline
0 & 0 & 0 & 1
\end{array}
\right]
\left\{
\begin{array}{c}
Y_1 \\ \hline Y_2 \\ \hline Y_3 \\ \hline Y_4
\end{array}
\right\}
=
\left\{
\begin{array}{c}
\tilde{Y}_1 \\ \hline F_2 \\ \hline F_3 \\ \hline \tilde{Y}_4
\end{array}
\right\}
$$

Here again, for our particular problem the known values on the right-hand side are zero; hence, the solution of the above equation will give values for the assumed unknown values as zero. The advantage of this method is that there is no need to eliminate rows and columns in our stiffness matrix or eliminate them at the time the global stiffness matrix is created. Of course, a disadvantage is that we have kept known values in our set of equations as unknown values. However, the more important disadvantage is that the original symmetric matrix is now nonsymmetric (see the next section), and this change will require a significant increase in the solution time to solve the equations. Hence, we consider our final method.

Method 3. This method accomplishes the same thing as the second method without destroying the symmetry of our matrix. We simply multiply the diagonal term of each of the two equations representing the known deflections by a very large number and replace the corresponding right-hand sides by the known deflections multiplied by the corresponding new diagonal. Thus, we have

$$
\begin{bmatrix}
\dfrac{T}{L_1} \times B & -\dfrac{T}{L_1} & 0 & 0 \\[2ex]
-\dfrac{T}{L_1} & \dfrac{T}{L_1} + \dfrac{T}{L_2} & -\dfrac{T}{L_2} & 0 \\[2ex]
0 & -\dfrac{T}{L_2} & \dfrac{T}{L_2} + \dfrac{T}{L_3} & -\dfrac{T}{L_3} \\[2ex]
0 & 0 & -\dfrac{T}{L_3} & \dfrac{T}{L_3} \times B
\end{bmatrix}
\begin{Bmatrix}
Y_1 \\[2ex] Y_2 \\[2ex] Y_3 \\[2ex] Y_4
\end{Bmatrix}
=
\begin{Bmatrix}
(\tilde{Y}_1)\dfrac{T}{L_1} \times B \\[2ex]
F_2 \\[2ex]
F_3 \\[2ex]
(\tilde{Y}_4)\dfrac{T}{L_3} \times B
\end{Bmatrix}
\tag{3.25}
$$

where B is a number several orders of magnitude larger than any other term in the matrix. The best way to understand how this works is to divide the first (or last) equation by its diagonal term. This gives a diagonal equal to unity and makes the off-diagonal term equal to $(-1/B)$. The right-hand side becomes $(\tilde{Y}_1)$. Thus, because $(1/B)$ is very small, we essentially have the same equation we created for our second method. We have succeeded, therefore, in forcing the two unwanted equations to reflect our boundary conditions while keeping the symmetry of our matrix, and we have not had to rearrange the matrix to accommodate this change. The disadvantage of this method is that we are solving a larger set of equations than we actually need to solve; that is, we still have our equations representing the known values. However, the ease with which this method is implemented in a computer code has made it very popular. We will discover later that it allows us to easily change boundary conditions during analyses of time-dependent problems and during the solution of nonlinear problems where the boundary conditions depend on the final solution. This procedure will be the standard procedure used in the programs presented in this text and will be referred to as "blasting" the diagonal.

3.6 COMPACT STORAGE OF THE STIFFNESS MATRIX

If we examine the stiffness matrix in Eq. 3.25, we will notice two important characteristics. The first is that $K(I, J) = K(J, I)$. For example, the third term (entry) on row 2 is equal to the second term on row 3. Such matrices are referred to as *symmetric* matrices. The second characteristic is that the nonzero terms all appear close to the diagonal. Such matrices are referred to as *banded* matrices. The maximum number of terms included between the first nonzero term and the last nonzero term on any row is referred to as the *bandwidth* of that row. Different rows of a matrix can have different bandwidths. The maximum bandwidth of all the rows is the bandwidth of the matrix. Hence, for our example, the bandwidth is 3. We will see in later chapters that these characteristics are true for a large number of finite element formulations.

These characteristics are important because they allow us to store the matrix in a compact form. In addition, this compact storage arrangement allows a much faster solution time for the matrix equation. To illustrate how we will store our stiffness matrix, we return to our example problem and its stiffness matrix as shown in Eq. 3.25. Because this matrix is symmetric, there is no need to store those terms "beneath" the diagonal. For example, we do not need to store $K(3, 2)$ because we know its value is equal to $K(2, 3)$. Thus, we eliminate all these terms and store only

$\frac{T}{L_1} \times B$	$-\frac{T}{L_1}$	0	0
	$\frac{T}{L_1} + \frac{T}{L_2}$	$-\frac{T}{L_2}$	0
		$\frac{T}{L_2} + \frac{T}{L_3}$	$-\frac{T}{L_3}$
			$\frac{T}{L_3} \times B$

Next, because we know that all terms outside the bandwidth are zero, we do not need to store them:

$\frac{T}{L_1} \times B$	$-\frac{T}{L_1}$		
	$\frac{T}{L_1} + \frac{T}{L_2}$	$-\frac{T}{L_2}$	
		$\frac{T}{L_2} + \frac{T}{L_3}$	$-\frac{T}{L_3}$
			$\frac{T}{L_3} \times B$

Now, because there are only two terms per row that we need to store, we use the following matrix:

$\dfrac{T}{L_1} \times B$	$-\dfrac{T}{L_1}$
$\dfrac{T}{L_1} + \dfrac{T}{L_2}$	$-\dfrac{T}{L_2}$
$\dfrac{T}{L_2} + \dfrac{T}{L_3}$	$-\dfrac{T}{L_3}$
$\dfrac{T}{L_3} \times B$	

In this final form, the first column of any row is the diagonal term. This compact storage has one space that is not needed; the second column of the last row. It represents a term that would actually be outside the original matrix. However, the savings in space obtained through this arrangement certainly compensates for having to create some fictitious locations. To fully appreciate the savings in storage, consider that if we had 100 nodes, we would need only 200 storage locations as compared with 10,000 if the entire matrix were saved. Because the matrix is symmetric, the number of columns needed to store the nonzero terms has been reduced to 2. We thus define the bandwidth for this symmetric matrix as 2.

Of course, if we store our stiffness matrix in this form, we must have an *equation solver* that recognizes this type of storage and can solve the matrix without having to expand it to its full size. To do this, the solver must be given not only the compact stiffness matrix and the corresponding right-hand side, but also the number of rows (number of equations) and the number of columns (the bandwidth) of the stiffness matrix. The particular solver used is one based on Gauss elimination for symmetric, banded matrices and is named sGAUSS.m It is described in Appendix A.

3.7 PROGRAM wire.m

We now present a complete finite element code that solves our tight wire problem. It is a prototype of all finite element codes presented in this text; hence, careful study of the code will help you understand the structure of the more complex codes to follow. The principal parts of the code and the definition of the variables are as follows:

Input Data

NUMEL	The number of elements.
TENS	The tension in the wire.
XORD(I)	The array of nodal point coordinates.
F(I)	The array of nodal point forces.
NPBC(I)	Array of nodal point boundary conditions.
	If NPBC(I) = 0, deflection at node I is unknown.
	If NPBC(I) = 1, deflection is known.

U(I)   The deflection of node *I* if known, otherwise this value is not used.

Initialization. This is a standard section in FEM programs and is used to initialize all parameters and arrays that are not read in as INPUT. For the tight wire problem,

NUMNP = NUMEL+1   The number of nodal points.

SK(I,J) = 0   The stiffness matrix.

IB = 2   The bandwidth.

Formation of SK Matrix. This section creates the stiffness matrix and is the heart of most finite element programs. In our case, it is quite simple because each element of the stiffness matrix is simply $\pm T/L_e$.

Boundary Conditions. This section searches through the NPBC array to determine which nodes have specified deflections. For each of these nodes, the specified deflection is enforced by decoupling these equations through the technique of blasting the diagonal.

Call Equation Solver to Determine Solution. Once the stiffness matrix has been formed and the boundary conditions specified, we need to solve the matrix equation for the unknown deflections. This is done through a call to an equation solver. In our case, this will be **sGAUSS.m**. Appendix A gives the details of this solver.

Output Data. The final part of any FEM program is the output of results. Most FEM programs have large amounts of output; hence, a major concern of FEM workers is how to present this output in a manner that is useful and easy to interpret. For the tight wire problem, the deflections are presented in graphical form. Both the nodal point coordinates and their deflections are saved in the working directory as XORD and U.

Flow Chart

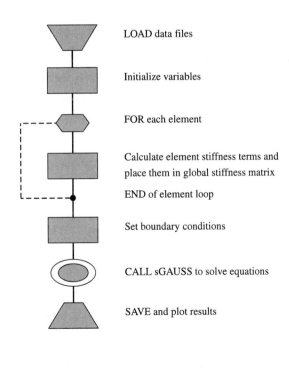

LOAD data files

Initialize variables

FOR each element

Calculate element stiffness terms and place them in global stiffness matrix

END of element loop

Set boundary conditions

CALL sGAUSS to solve equations

SAVE and plot results

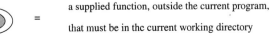

= a supplied function, outside the current program, that must be in the current working directory

Code

```
  clear
% -------------------
% INPUT DATA
% -------------------
  load MSHDAT   -ASCII
  load NODES    -ASCII

  TENS  = MSHDAT(1);
  NUMEL = MSHDAT(2);

  NUMNP=NUMEL+1;

  for  I=1:NUMNP
      XORD(I)=NODES(I,1);
      NPBC(I)=NODES(I,2);
      U(I)   =NODES(I,3);
      F(I)   =NODES(I,4);
  end


% --------------
% INITIALIZATION
% --------------
  IB=2;
  for I=1:NUMNP;
      for  J=1:IB;
          SK(I,J)=0.0 ;
      end;
  end;

% ----------------------------
% FORMATION OF STIFFNESS MATRIX
% ----------------------------
  for I=1:NUMEL;
      RL=XORD(I+1)-XORD(I);
      RK=TENS/RL;
      SK(I,1)=SK(I,1)+RK;
      SK(I,2)=SK(I,2)-RK;
      IP1=I+1;
      SK(IP1,1)=SK(IP1,1)+RK;
  end;
```

Load mesh data and quadrature data
Both files must be in working directory

TENS	Tension in wire
NUMEL	Number of elements
NUMNP	Number of nodal points
XORD	x coordinates of nodes
NPBC	Nodal point boundary condition
	NPBC = 0: deflection unknown
	NPBC = 1: deflection known
U	Deflections in: $[K]\{U\} = \{F\}$
F	Forces in: $[K]\{U\} = \{F\}$
IB	Bandwidth of stiffness matrix
SK	Global stiffness matrix, $SK = K$ in $[K]\{U\} = \{F\}$

Begin loop over all elements to create
element stiffness matrix.

I	Current element number (also global number of left node "a")
RL	Length of current element
RK	T/L_e for current element
SK	Compacted global stiffness matrix
XI	Coordinate of left node of current element
IP1	Global number of right node "b"

```
% --------------------
% BOUNDARY CONDITIONS
% --------------------
  for I=1:NUMNP;
     if NPBC(I) == 1;
        SK(I,1)=SK(I,1)*1.0E+06;
        F(I)=U(I)*SK(I,1);
     end;
  end;




% ---------------------------------
% CALL EQUATION SOLVER
% ---------------------------------
  U =  sGAUSS(SK,F,NUMNP,IB);





% -----------
% OUTPUT DATA
% -----------
  save XORD XORD -ASCII
  save U U -ASCII

  plot(XORD,U)
  grid
  xlabel(' Distance along wire ')
  ylabel(' Deflection of wire   ')
  title('[Deflection of a tightly'...
        ' stretched wire'])
```

I	Node number
if	NPBC(I) = 1, then deflection is known at node *I*
SK(I,1)	Stiffness diagonal on row *I*
1.0E+06	Blasting term to decouple row *I*
F(I)	Right-hand side for row *I*
U(I)	Specified value for *U* obtained from data file NODES
sGAUSS	Equation solver for banded, symmetric matrices. Code must be in working directory. Solves: $[SK]\{U\} = \{F\}$ (see Appendix A)
IB	Bandwidth. Set equal to 2 at beginning of code.
U	Wire deflections
save	MATLAB routine to save variables in ASCII format
plot	MATLAB routine Plots *U* vs. *x*

Test Problem. It is always wise, after writing a finite element code, to test it using a very simple problem. Often these problems are chosen so that the code will duplicate the exact answer. Sometimes, as in the following example, it is sufficient to use a problem for which we know the exact answer though the code may not necessarily duplicate it. In either case, the idea is to check the code using very few elements so that, if necessary, numerical errors can be traced without the creation of large output files.

This test problem involves a wire of length 12 units, with a tension of 20,000 units of force and a uniform load of 1 unit of force per unit of length. Two data files are used to describe the problem, MSHDAT and NODES. They are described in the discussion that follows. It is necessary to have in your working directory, along with these data files and wire.m, the equation solver, sGAUSS.m. This solver can be found in Appendix A.

MSHDAT

```
% tension    number of elements
%----------------------------------
   20000.0              8
%----------------------------------
```

NODES

```
%  XORD    NPBC      U       F
%----------------------------------
   0.0       1       0     -1.0
   2.0       0       0     -2.0
   4.0       0       0     -1.5
   5.0       0       0     -1.0
   6.0       0       0     -1.0
   7.0       0       0     -1.0
   8.0       0       0     -1.5
  10.0       0       0     -2.0
  12.0       1       0     -1.0
%----------------------------------
```

For the most part, these two files are self-explanatory. However, you should note the following:

1. The coordinate of the first node is 0.0; hence, the length of the wire equals the coordinate of the last node. However, it is not necessary that the first node be at the origin of the coordinate system used. The length of the wire is always the difference between the first and last nodal coordinates.

2. The loads shown for each node represent those obtained from the integration of

$$\int_0^{L_e} \{N\} w \, dx$$

where $\{N\}$ is the column of the two shape functions, and w was taken as a constant unit loading per unit length. The total load for any given element is, therefore, the length of the element. Half of this load went to each of the element's two nodes.

3. For the first and last nodes, the half loadings have been included. However, these values are neither used nor needed because the deflections are known at these points as indicated by their NPBC value.

4. For these same two boundary nodes, U was specified as zero. However, you may specify any value you wish and wire.m will enforce that boundary condition. In fact, you can specify additional known deflections at nodes between the end points, e.g., a support at midspan that holds the wire up by a given amount.

If the program is run using the above data, the following plot will be generated:

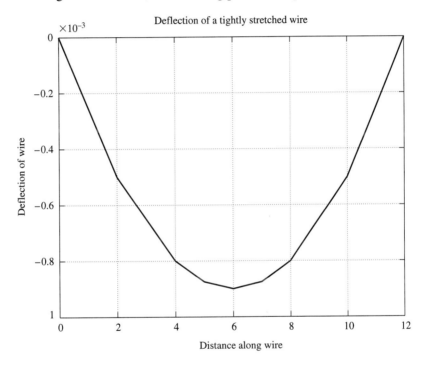

EXERCISES

Study Problems

S1. Using the general equation for F_a and F_b, determine their values when a concentrated load P is placed a distance z from the left-hand end of the element. (*Hint:* Treat P as the resultant of a uniform distributed load w over a distance Δx at $x = z$, and then let Δx approach zero.)

S2. Determine F_a and F_b for a linearly distributed load that varies from w_a at the left-hand end to w_b at the right-hand end of the element.

S3. Consider a single element of length 2.0 m with a loading

$$w(x) = 10.0 - 6.0x + 5.0x^2 \text{ N/m}$$

Calculate F_a and F_b using Eqs. 3.12.

S4. Place the following matrix in the compact storage arrangement for use with the sGAUSS.m equation solver. What is its bandwidth?

6	−2	−2	0	0	0	0
−2	8	−3	−2	0	0	0
−2	−3	−7	−1	−2	0	0
0	−2	−1	8	−3	−1	0
0	0	−2	−3	6	−1	−2
0	0	0	−1	−1	8	−3
0	0	0	0	−2	−3	7

S5. Expand the following matrix, given in compact storage for a banded symmetric matrix, to a full, square matrix.

8	−2	−3
7	−1	0
9	0	−1
5	−3	0
7	−3	−2
6	−1	
8		

S6. Show that F_a and F_b as given by Eqs. 3.12 correspond to the reactions of a simply supported beam under the same loading.

Numerical Experiments and Code Development

N1. Change the data given for the example problem to specify that the first node has a displacement of +0.001 units. Run wire.m for this condition.

N2. For the example problem, change the number of elements to be 12 with evenly spaced nodes. Run wire.m for this condition.

N3. Determine a finite element approximate solution for the deflection of the wire under the loading shown.

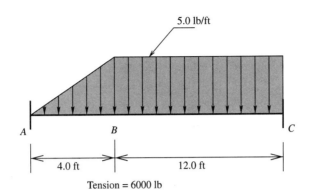

5.0 lb/ft

A   B   C

4.0 ft   12.0 ft

Tension = 6000 lb

Use five elements, all of different lengths, and plot the finite element solution with the exact solution.

N4. Write a program to automatically generate the files MSHDAT and NODES, with uniformly spaced nodes. Enter as data: length, tension, and number of elements. Also, have the program calculate the nodal forces from a user-written loading function for $w(x)$. Assume the loading is linear between nodes and use the equations developed in Study Problem 2.

N5. Same as Problem 3, but specify a support 6 feet to the right of A that holds the wire 0.005 feet above the position shown. Use the mesh generator you developed in Problem 4 to create a mesh of 32 elements. (You do not need to calculate and plot the exact solution.)

N6. Create a 12 × 4 matrix of your choice to represent a symmetric, banded matrix, $[A]$. Assume all entries of $\{x\}$ equal to unity, and calculate $\{F\}$, where

$$[A]\{x\} = \{F\}$$

Write a short program that uses sGAUSS.m to solve the above for $\{x\}$ for your calculated values of $\{F\}$.

N7. Repeat the preceding exercise, but specify one of the values as known. Use the blasting technique as described in the text and used in wire.m.

N8. Use a wire of unit length, a constant uniform load w_o, a tension T, and two elements with a node in the center of the wire to numerically demonstrate that the deflection is a linear function of $(1/T)$ and w. Do this by running two sets of experiments:
(a) For $T = 1.0$, let $w_o = 1, 2,$ and 4, and plot nodal deflection versus w_o.
(b) For $w_o = 1.0$, let $T = 1, 2,$ and 5, and plot nodal deflection versus $(1/T)$.
    Explain why the outcome was already known based on the governing equation and the finite element approximation.

4

LINEAR SECOND-ORDER ORDINARY DIFFERENTIAL EQUATIONS

In this chapter we use the finite element method to obtain approximate solutions to second-order linear ordinary differential equations[1] of the form

$$\frac{d}{dx}\left[A(x)\frac{dy}{dx}\right] + B(x)\frac{dy}{dx} + C(x)y + D(x) = 0.0 \tag{4.1}$$

with the boundary conditions

$$\text{at } x = 0 \quad y = y_0^* \quad \text{or} \quad A\frac{dy}{dx} = -q_0^*$$

$$\text{at } x = L \quad y = y_L^* \quad \text{or} \quad A\frac{dy}{dx} = q_L^* \tag{4.2}$$

In order to determine a unique solution to Eq. 4.1, it is necessary to specify one of the two boundary conditions shown at $x = 0$ and at $x = L$. For problems where $C(x)$ is identically zero, it is necessary that y be one of the specifications at either $x = 0$ or $x = L$; the boundary condition at the remaining end can be specified with either of the two conditions. For most applications, the second type of boundary condition represents a flux, either into or out of the region. The direction of the flux is usually opposite that of the positive gradient of the dependent variable y. It is common to designate such a flux as positive when it represents a flow into the region—hence the negative sign in front of q_0^* in Eqs. 4.2.

4.1 WEAK FORM OF THE GOVERNING EQUATION

We now place the preceding equation in its weak form, suitable for the finite element method. We will designate the test functions as δy to emphasize the relationship between the weak form and the variational formulation. Thus,

$$\int_0^L \delta y \left[\frac{d}{dx}\left(A\frac{dy}{dx}\right) + B\frac{dy}{dx} + Cy + D\right] dx = 0 \tag{4.3}$$

---

[1]This chapter is limited to the analysis of second-order differential equations. Chapter 13 covers higher-order ordinary differential equations and is suitable for study immediately following this chapter. The treatment of higher-order partial differential equations by the finite element method requires considerations that are beyond the scope of this introductory text.

where the function y represents the exact solution (the function that satisfies Eqs. 4.1 and 4.2) and δy represents an arbitrary function that we interpret as a variation in y.

Before proceeding, we note the similarity between Eq. 4.3 and Galerkin's method. If a family of approximate solutions were used for y and the same family of approximations were used for δy, the solution of Eq. 4.3, for all possible independent functions δy, would be the solution to our problem by Galerkin's method. However, this would require that the second derivatives of our approximating functions be defined at all points. This is not the case for our finite element approximations; hence, we integrate by parts to eliminate the second-derivative term. This will place a weaker demand on our approximating functions in terms of smoothness.

Thus, we begin our integration by parts:

$$\delta y \frac{d}{dx}\left[A\frac{dy}{dx}\right] = \frac{d}{dx}\left[\delta y A \frac{dy}{dx}\right] - \frac{d\delta y}{dx}A\frac{dy}{dx} \tag{4.4}$$

which, when substituted into the integral, gives us

$$\int_0^L \delta y \frac{d}{dx}\left[A\frac{dy}{dx}\right]dx = \int_0^L \frac{d}{dx}\left[\delta y A \frac{dy}{dx}\right]dx - \int_0^L \frac{d\delta y}{dx}A\frac{dy}{dx}dx \tag{4.5}$$

These steps have placed an additional restriction on our function δy. Not only must it be defined at each point in the domain, but its derivative must be defined; hence, δy must be a continuous function.

The first term on the right-hand side of Eq. 4.5 can be integrated directly provided the term in the brackets is a continuous function. Thus,

$$\int_0^L \delta y \frac{d}{dx}\left[A\frac{dy}{dx}\right]dx = \left.\delta y A \frac{dy}{dx}\right|_0^L - \int_0^L \frac{d\delta y}{dx}A\frac{dy}{dx}dx \tag{4.6}$$

Substitution of Eq. 4.6 into Eq. 4.3 gives

$$\int_0^L \left[-\frac{d\delta y}{dx}A\frac{dy}{dx} + \delta y B \frac{dy}{dx} + \delta y C y + \delta y D\right]dx + \left.\delta y A \frac{dy}{dx}\right|_0^L = 0 \tag{4.7}$$

It is important to consider the last term in this equation. It appears as a result of our integration by parts and involves the boundary points at $x = 0$ and $x = L$. If we use Eqs. 4.2, we can write

$$\begin{aligned}
\left.\delta y A \frac{dy}{dx}\right|_0^L &= \left.\delta y_L A_L \frac{dy}{dx}\right|_L - \left.\delta y_0 A_0 \frac{dy}{dx}\right|_0 \\
&= \delta y_L q_L \qquad\quad - \delta y_0(-q_0) \\
&= \delta y_L q_L \qquad\quad + \delta y_0 q_0
\end{aligned} \tag{4.8}$$

Either y or q must be specified at each of the two endpoints. If y is known, then we will limit the trial functions to those that have this value and require the test functions, δy, to be zero at that point. If we do not know the value for y, then we must know q. We write Eq. 4.7 as

$$\int_0^L \left[-\frac{d\delta y}{dx} A \frac{dy}{dx} + \delta y B \frac{dy}{dx} + \delta y C y + \delta y D \right] dx + \delta y_0 q_0 + \delta y_L q_L = 0 \tag{4.9}$$

We have now found that the solution to Eq. 4.1 that satisfies the boundary conditions expressed by Eqs. 4.2, satisfies Eq. 4.9 for all continuous test functions δy. If a known value of either q_0 or q_L is specified, then the equation is satisfied for arbitrary values of δy at that point. If either of these values is unknown, then the equation is satisfied for any δy that has a zero value at the location corresponding to the unknown q. Furthermore, if we follow the above steps backward, we will find that, out of the family of all continuous trial functions, there is only one function that satisfies Eq. 4.9 for all continuous test functions δy, and that function is the one that satisfies Eq. 4.1 with boundary conditions expressed by Eq. 4.2.

We need to ask what will happen if we confine both the trial and the test functions to our piecewise linear, continuous FEM family, that is, if we exclude functions with continuous first derivatives including the exact solution. Is there a finite element approximation that satisfies Eq. 4.9 for every δy from this same family of approximations? The answer is yes, there is one–and only one. It will be the solution obtained by solving our set of finite element linear algebraic equations. We note, however, that this solution might not satisfy Eq. 4.9 for test functions δy outside this set; therefore, it is not necessarily the exact solution. However, it does seem reasonable that the best approximation from the FEM set of functions be defined as the one that satisfies Eq. 4.9 for all test functions from this same set. With this understanding, we proceed with our finite element formulation.

4.2 FINITE ELEMENT FORMULATION

Equation 4.9 is best expressed as the sum of four separate integrals. Their integration will be done element by element and will lead to three separate element stiffness matrices and a right-hand-side forcing vector. Each of the stiffness matrices contributes to a global stiffness matrix. The global stiffness matrix and the global forcing vector make up our FEM set of simultaneous equations.

We will consider a typical element and write all four of the integrals in our shorthand notation. Once they are written in this general form, we will expand the terms for use in writing our finite element program.

First, we review our notation:

$$y = \lfloor N \rfloor \{Y\} \tag{4.10}$$

$$\frac{dy}{dx} = \lfloor dN/dx \rfloor \{Y\}$$
$$= \lfloor N' \rfloor \{Y\} \tag{4.11}$$

$$\delta y = \lfloor N \rfloor \{\delta Y\} \tag{4.12}$$

$$\delta \frac{dy}{dx} = \lfloor dN/dx \rfloor \{\delta Y\}$$
$$= \lfloor N' \rfloor \{\delta Y\} \tag{4.13}$$

With this notation, the four integrals associated with Eq. 4.9 are

$$\int_0^{L_e} \{N'\}A\lfloor N'\rfloor \; dx = [S_1] \tag{4.14}$$

$$\int_0^{L_e} \{N\}B\lfloor N'\rfloor \; dx = [S_2] \tag{4.15}$$

$$\int_0^{L_e} \{N\}C\lfloor N\rfloor \; dx = [S_3] \tag{4.16}$$

$$\int_0^{L_e} \{N\}D \; dx = \{f\} \tag{4.17}$$

where we have used the notation $\{\cdot\} = \lfloor\cdot\rfloor^T$.

Addition of these matrices gives us the total contribution to the integral in Eq 4.9 from element e:

$$\delta J_e = -\lfloor \delta Y \rfloor [S_1]\{Y\} + \lfloor \delta Y \rfloor [S_2]\{Y\} + \lfloor \delta Y \rfloor [S_3]\{Y\} + \lfloor \delta Y \rfloor \{f\} \tag{4.18}$$

where we have used the symbol δJ_e to correspond to a variation of a quantity that might or might not have a corresponding functional. This is similar to using such a symbol to represent a variation in work due to a variation in displacement even though there is no corresponding potential energy functional. It likewise corresponds to a differential in multivariable calculus, whether exact or inexact. The notation is convenient and should not be taken to indicate that a corresponding functional, J, does exist.

After the contributions to δJ from each element have been added to the global stiffness matrix and to the global forcing vector, $\{F\}$, it is necessary to add the contributions to the forcing vector from the two boundary points as indicated in Eq. 4.9. After all contributions have been accounted for, we have

$$\sum_1^{\text{numel}} \left(\lfloor \delta Y \rfloor \Big[[S_1] - [S_2] - [S_3] \Big]\{Y\} - \lfloor \delta Y \rfloor \{F\} \right) - \delta Y_0 q_0 - \delta Y_L q_L = 0 \tag{4.19}$$

Note the change in signs for all terms. This is customary so as to associate a positive sign with the $[S_1]$ contribution—more appropriate for many physical interpretations.

The last two terms represent the boundary conditions at $x = 0$ and $x = L$ and can be added to the forcing vector to give

$$\{F\} = \begin{Bmatrix} F_1 + q_0 \\ F_2 \\ \vdots \\ F_{n-1} \\ F_n + q_L \end{Bmatrix}$$

Thus, Eq. 4.19 can be written

$$\delta J = \lfloor \delta Y \rfloor [K] \{Y\} - \lfloor \delta Y \rfloor \{F\} = 0 \tag{4.20}$$

Here $[K]$ represents the total global stiffness matrix made up of all three separate matrices shown in Eq. 4.19.

At this point in the development for the tightly stretched wire we stated that δY at the end nodes had to be zero for the approximating function to satisfy the specified boundary conditions. However, that is not the case for our current formulation. Because we have included the end fluxes, q_0 and q_L, in the above equation, it is valid for all δY. This would also have been true for our wire problem had we included in $\{F\}$ the reaction forces at each end of the wire.

The equation is therefore valid regardless of which boundary condition is specified, and we need not, at this point, decide which will be specified. Thus, Eq. 4.20 must be satisfied for all $\{\delta Y\}$, and this requires

$$[K]\{Y\} - \{F\} = 0 \tag{4.21}$$

Note, however, that although this equation is valid, the stiffness matrix might be singular; thus, a unique solution might not exist. To remove this singularity, it is necessary to specify at least one nodal value of Y.

The matrices $[S_1]$, $[S_2]$, $[S_3]$ and the vector $\{F\}$, when expanded, are

$$[S_1] = \int_0^{L_e} \{N'\} \; A(x) \; \lfloor N' \rfloor dx$$

$$= \int_0^{L_e} \begin{Bmatrix} N_L' \\ N_R' \end{Bmatrix} \; A(x) \; \lfloor N_L' \quad N_R' \rfloor dx \tag{4.22}$$

$$= \int_0^{L_e} \begin{bmatrix} N_L' A N_L' & N_L' A N_R' \\ N_R' A N_L' & N_R' A N_R' \end{bmatrix} dx$$

$$[S_2] = \int_0^{L_e} \{N\} \; B(x) \; \lfloor N' \rfloor dx$$

$$= \int_0^{L_e} \begin{Bmatrix} N_L \\ N_R \end{Bmatrix} \; B(x) \; \lfloor N_L' \quad N_R' \rfloor dx \tag{4.23}$$

$$= \int_0^{L_e} \begin{bmatrix} N_L B N_L' & N_L B N_R' \\ N_R B N_L' & N_R B N_R' \end{bmatrix} dx$$

$$[S_3] = \int_0^{L_e} \{N\} \; C(x) \; \lfloor N \rfloor dx$$

$$= \int_0^{L_e} \begin{Bmatrix} N_L \\ N_R \end{Bmatrix} \; C(x) \; \begin{bmatrix} N_L & N_R \end{bmatrix} dx \qquad (4.24)$$

$$= \int_0^{L_e} \begin{bmatrix} N_L C N_L & N_L C N_R \\ N_R C N_L & N_R C N_R \end{bmatrix} dx$$

$$[f] = \int_0^{L_e} \{N\} \; D(x) \; dx$$

$$= \int_0^{L_e} \begin{Bmatrix} N_L \\ N_R \end{Bmatrix} \; D(x) \; dx \qquad (4.25)$$

$$= \int_0^{L_e} \begin{bmatrix} N_L D \\ N_R D \end{bmatrix} dx$$

4.3 CONSTANT COEFFICIENTS

In order to carry out the integrations indicated, the functions $A(x)$, $B(x)$, $C(x)$, and $D(x)$ must be specified. In the next section a method for performing the integration within the finite element program will be described. For now, let us perform the integrations for the special case when the parameters have constant values within an element. Such would be the case if they were constant throughout the entire domain, or simply piecewise constant through the domain. It would also be the case for coefficients that were functions of x if the elements were taken so small that within each, the coefficients could be considered constant and the average value used.

The first integral is exactly what we used for the tight wire problem and is equal to

$$[S_1] = \begin{bmatrix} \dfrac{A}{L_e} & -\dfrac{A}{L_e} \\[2ex] -\dfrac{A}{L_e} & \dfrac{A}{L_e} \end{bmatrix} \qquad (4.26)$$

To obtain the second integral we must integrate each of the functions $S_2(1, 1)$, $S_2(1, 2)$, $S_2(2, 1)$, and $S_2(2, 2)$. We have for the first term

$$S_2(1, 1) = \int_0^{L_e} N_L B N'_L dx$$

$$= \int_0^{L_e} \left(1 - \frac{x}{L_e}\right) B \left(-\frac{1}{L_e}\right) dx$$

$$= -B \frac{1}{L_e} \int_0^{L_e} \left(1 - \frac{x}{L_e}\right) dx$$

$$= -B \frac{1}{L_e} \left(x - \frac{x^2}{2L_e}\right) \Big|_0^{L_e} \tag{4.27}$$

$$= -B \frac{1}{L_e} \left(L_e - \frac{1}{2}L_e\right)$$

$$= -\left(\frac{B}{2}\right)$$

The second term differs from $S_2(1, 1)$ only by its sign, and is

$$S_2(1, 2) = \int_0^{L_e} N_L B N'_R dx$$

$$= B \int_0^{L_e} \left(1 - \frac{x}{L_e}\right) \left(+\frac{1}{L_e}\right) dx \tag{4.28}$$

$$= +\frac{B}{2}$$

The third term is

$$S_2(2, 1) = \int_0^{L_e} N_R B N'_L dx$$

$$= B \int_0^{L_e} \left(\frac{x}{L_e}\right) \left(-\frac{1}{L_e}\right) dx \tag{4.29}$$

$$= -\frac{B}{2}$$

Note that $S_2(1, 2)$ and $S_2(2, 1)$ are not equal; hence, this matrix is not symmetric. The fourth and last term is

$$S_2(2, 2) = \int_0^{L_e} N_R B N_R' \, dx$$

$$= B \int_0^{L_e} \left(\frac{x}{L_e}\right)\left(\frac{1}{L_e}\right) dx \tag{4.30}$$

$$= +\frac{B}{2}$$

The full matrix can now be written as

$$[S_2] = \begin{bmatrix} -\dfrac{B}{2} & \dfrac{B}{2} \\[2mm] -\dfrac{B}{2} & \dfrac{B}{2} \end{bmatrix} \tag{4.31}$$

Next we consider the integration of

$$[S_3] = \int_0^{L_e} \{N\} C \lfloor N \rfloor \, dx \tag{4.32}$$

This integral contains the square of the shape functions and therefore squares of linear functions. Hence, the integrands of the four terms of the matrix $[S_3]$ will be quadratic in x. The first term is

$$S_3(1, 1) = \int_0^{L_e} N_L C N_L \, dx$$

$$= \int_0^{L_e} \left(1 - \frac{x}{L_e}\right) C \left(1 - \frac{x}{L_e}\right) dx$$

$$= C \int_0^{L_e} \left(1 - 2\frac{x}{L_e} + \frac{x^2}{L_e^2}\right) dx \tag{4.33}$$

$$= C \left(L_e - L_e + \tfrac{1}{3}L_e\right)$$

$$= \tfrac{1}{3} C L_e$$

In a similar way, the other three terms in this matrix can be integrated to give

$$[S_3] = \begin{bmatrix} \dfrac{CL_e}{3} & \dfrac{CL_e}{6} \\ \dfrac{CL_e}{6} & \dfrac{CL_e}{3} \end{bmatrix} \tag{4.34}$$

We now have all of the integrals evaluated that contribute to the global stiffness matrix. The right-hand side is evaluated using Eq. 4.26. For constant D the element force vector is

$$\left\{ \begin{array}{c} f_L \\ f_R \end{array} \right\} = \left\{ \begin{array}{c} \dfrac{DL_e}{2} \\ \dfrac{DL_e}{2} \end{array} \right\} \tag{4.35}$$

The sum of all matrices for a single element is

$$[k]_e = \begin{bmatrix} \left(\dfrac{A}{L_e} + \dfrac{B}{2} - \dfrac{CL_e}{3} \right) & \left(-\dfrac{A}{L_e} - \dfrac{B}{2} - \dfrac{CL_e}{6} \right) \\ \left(-\dfrac{A}{L_e} + \dfrac{B}{2} - \dfrac{CL_e}{6} \right) & \left(\dfrac{A}{L_e} - \dfrac{B}{2} - \dfrac{CL_e}{3} \right) \end{bmatrix} \tag{4.36}$$

and

$$\{f\}_e = \left\{ \begin{array}{c} \dfrac{DL_e}{2} \\ \dfrac{DL_e}{2} \end{array} \right\} \tag{4.37}$$

Assembly of these terms into the global stiffness matrix and the global forcing vector leads to the set of algebraic equations defining the finite element approximation of the solution of Eq. 4.1.

4.3.1 Illustrations

Tight Wire on Elastic Foundation. Figure 4.1 illustrates our tight wire problem with the addition that it is resting on an elastic foundation that exerts a force proportional in magnitude and opposite in direction to the deflection of the wire. The differential equation for the deflection of the wire is now

$$\frac{d}{dx}\left[T\frac{dy}{dx} \right] - ky + w = 0 \tag{4.38}$$

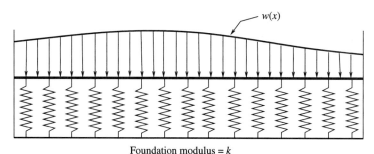

Foundation modulus = k

Figure 4.1.   Wire on an elastic foundation.

A comparison of this equation with our general second-order differential equation, Eq. 4.1, shows that the coefficient A is the tension T, the coefficient C is the negative of the foundation modulus k, and the coefficient D is the loading. The coefficient B is zero for this problem; therefore, the stiffness matrix will be symmetric and is

$$
\begin{bmatrix}
\left(+\dfrac{T}{L_e} + \dfrac{kL_e}{3} \right) & \left(-\dfrac{T}{L_e} + \dfrac{kL_e}{6} \right) \\[3mm]
\left(-\dfrac{T}{L_e} + \dfrac{kL_e}{6} \right) & \left(+\dfrac{T}{L_e} + \dfrac{kL_e}{3} \right)
\end{bmatrix}
\tag{4.39}
$$

The force vector for a constant load is

$$
\left\{
\begin{matrix}
f_L \\[3mm]
f_R
\end{matrix}
\right\}
=
\left\{
\begin{matrix}
\dfrac{w_0 L_e}{2} \\[3mm]
\dfrac{w_0 L_e}{2}
\end{matrix}
\right\}
\tag{4.40}
$$

We see that the program wire.m could easily be modified to accommodate elastic foundations by adding the value of k to the input file and including the (kL_e) terms in the element stiffness matrix. The global stiffness matrix would remain symmetric; thus, no other modification would be necessary.

Heat Transport in a Pipe.   As a second illustration we consider an insulated pipe with a fluid flowing through it with an average, steady velocity equal to v. Let ρ be the density of the fluid, k its thermal conductivity, and C_p its specific heat. Conservation of energy then requires that the the temperature distribution, Θ, along the length of the pipe satisfy

$$
\frac{d}{dx}\left(k\frac{d\Theta}{dx} \right) - \rho C_p v \frac{d\Theta}{dx} = 0
\tag{4.41}
$$

where the coefficients are average values at a cross-section.

The boundary conditions at the ends of the pipe can be either a specified heat flux due to conduction or a specified temperature; however, the temperature must be specified at at least one end. We note from Eq. 4.41 that in terms of the notation of Eq. 4.1, $A = k$ and $B = -\rho C_p v$. Our element matrix for this case is therefore

$$
\begin{bmatrix}
\left(+\dfrac{k}{L_e} - \dfrac{\rho C_p v}{2}\right) & \left(-\dfrac{k}{L_e} + \dfrac{\rho C_p v}{2}\right) \\[2ex]
\left(-\dfrac{k}{L_e} - \dfrac{\rho C_p v}{2}\right) & \left(+\dfrac{k}{L_e} + \dfrac{\rho C_p v}{2}\right)
\end{bmatrix}
\tag{4.42}
$$

Although the two submatrices that make up this matrix are similar in form, there is an important difference that is best seen by considering that the elements are of equal length. If so, then when the element stiffness matrices are assembled into the global matrix, the k/L_e diagonal terms will add together, whereas the $\rho C_p v/2$ diagonal terms cancel. If not for the k/L_e terms, the diagonal terms of our matrix would all be zeros except at the two endpoints. When the $\rho C_p v/2$ terms become large, the off-diagonal terms can become larger in magnitude than the diagonal terms. Depending on the boundary conditions, this can create unwanted oscillations in the approximating function. However, because the element length does not appear in the convection term, the conduction term can always be made large in comparison by reducing the length of the element. Thus, we have a method of controlling the oscillations when they do appear simply by reducing the size of the elements.

4.4 VARIABLE COEFFICIENTS AND GAUSSIAN QUADRATURE

The formulation of our problem allows the coefficients A, B, C, and D to be functions of x. When this is the case, we will need to insert these functions into our code at the appropriate spot. In addition, we will have to numerically integrate the three stiffness matrices and the right-hand-side forcing vector. Because the numerical integrations will be performed for each element, it is important to select an efficient technique. The procedure used in almost all finite element codes is Gaussian quadrature.

Gaussian quadrature is a method of determining an integral by using a weighted average of the integrand evaluated at specified sampling points. The sampling points and the weights are given for an integration between the limits of -1 and $+1$. That is,

$$
\int_{-1}^{+1} f(u)\, du = \sum_{i=1}^{n} W_i f(u_i)
\tag{4.43}
$$

where n equals the number of sampling points, u_i are the coordinates of the points, and W_i are the weights. The weights and the coordinates are, of course, not arbitrary and must be available in a code. (See Appendix C for their derivation.)

It is shown in Appendix C that for n such points all polynomials up to degree $(2n - 1)$ are integrated exactly. Hence, we assume that for any integrand that can be accurately approximated with such a polynomial, Eq. 4.43 provides an accurate approximation for the integral. The following table gives the Gaussian coordinates and weights for $n = 1, 2,$ and 3.

n	Coordinates	Weights
1	0.0	2.0
2	−0.57735026918963	1.0
	+0.57735026918963	1.0
3	−0.77459666924148	0.55555555555556
	0.0	0.88888888888889
	+0.77459666924148	0.55555555555556

Consider now the integrands of the terms in $[S_3]$. If the coefficient, C, were constant, they would be second-degree polynomials, and two-point Gaussian quadrature would integrate them exactly. Integrands associated with $[S_2]$ and $[S_1]$ would be of lower degree if their coefficients were also constants. Hence, two points would be sufficient to integrate them exactly. Because it is always possible to select element sizes small enough that coefficients can be considered constant within their length, two-point Gaussian quadrature can always be used. However, it is sometimes computationally more efficient to use more quadrature points and fewer elements. At what point such a trade-off should be made is a mater of personal choice. Note that if three-point quadrature is used, the integrations will be accurate for integrands that can be approximated closely with a fifth-degree polynomial.

Because the limits of integration in Eq. 4.43 are ± 1, we must consider how to use this equation for other limits. To do so, it is only necessary to make a change in variables. Thus, if our integral is

$$\int_{x=a}^{b} f(x)\, dx \tag{4.44}$$

we can write x as the following function of u,

$$x = \left(\frac{a+b}{2}\right) + \left(\frac{b-a}{2}\right)u \tag{4.45}$$

and substitute this into our integral to obtain

$$\int_{x=a}^{b} f(x)\, dx = \int_{u=-1}^{+1} f[x(u)]\left(\frac{b-a}{2}\right) du \tag{4.46}$$

where

$$dx = \left(\frac{b-a}{2}\right) du \tag{4.47}$$

Thus, our integral is in a form applicable for Gaussian quadrature.

4.5 INTEGRATION OF THE ELEMENT MATRICES

Because the integrands of our element matrices have as factors the shape functions and their derivatives, and because these integrals will be evaluated by Gaussian quadrature, it is practical to write the shape functions in terms of the Gaussian coordinate u as shown in Fig. 4.2.

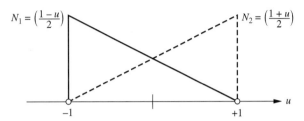

Figure 4.2. Shape functions in terms of the Gaussian coordinate.

The relationship between the Gaussian coordinate, u, and the global coordinate, x, has already been given by Eq. 4.45. We must now determine the relationship between dN/dx and dN/du. This is found by the chain rule of differentiation. Thus,

$$\frac{dN}{dx} = \frac{dN}{du}\frac{du}{dx} \tag{4.48}$$

where, from the relationships shown in Eq. 4.47,

$$\frac{du}{dx} = \frac{2}{b-a} = \frac{2}{L_e} \tag{4.49}$$

If we now use Eq. 4.46 to integrate

$$S_1(I, J) = \int_a^b \frac{dN_I}{dx} A \frac{dN_J}{dx}\, dx \tag{4.50}$$

$$S_2(I, J) = \int_a^b N_I B \frac{dN_J}{dx}\, dx \tag{4.51}$$

$$S_3(I, J) = \int_a^b N_I C N_J\, dx \tag{4.52}$$

$$Q_e(I) = \int_a^b N_I D\, dx \tag{4.53}$$

we have

$$S_1(I, J) = \int_{-1}^{+1} \frac{2}{L_e} \frac{dN_I}{du} A \frac{dN_J}{du} \frac{2}{L_e} \frac{L_e}{2} \, du \tag{4.54}$$

$$S_2(I, J) = \int_{-1}^{+1} N_I B \frac{dN_J}{du} \frac{2}{L_e} \frac{L_e}{2} \, du \tag{4.55}$$

$$S_3(I, J) = \int_{-1}^{+1} N_I C N_J \frac{L_2}{2} \, du \tag{4.56}$$

and

$$Q_e(I) = \int_{1}^{+1} N_I D \frac{L_e}{2} \, du \tag{4.57}$$

Rather than canceling the terms $(L_e/2)$ with $(2/L_e)$ terms, they are left as is because they represent two different mathematical quantities that will not always cancel in situations to be considered later in the text. The shape functions and their derivatives are now, of course, functions of u; whereas the parameters A, B, C, and D should be thought of as functions of the global coordinate x. For any given quadrature point u, they can be evaluated by first determining the corresponding global coordinate, x, using Eq. 4.45. Thus, the integration of our element matrices is reduced to the evaluation of rather simple integrands at each of the given quadrature points.

4.6 PROGRAMMING PRELIMINARIES

We are now in a position to write a finite element code for the approximate solution to a second-order linear differential equation with variable coefficients. Although this new code will have the same outline as program wire.m, we will introduce some new techniques that will also be used in the codes developed in later chapters.

4.6.1 Compact Storage of Stiffness Matrix. Because of the nonsymmetric $[S_2]$ matrix, we will have to store the lower, as well as the upper, triangular part of our stiffness matrix and use a routine for Gaussian elimination written for this type of storage arrangement. The routine is nGAUSS.m, and the storage arrangement used is similar to that used for sGAUSS.m. The following diagram illustrates how to compactly store a banded nonsymmetric matrix to be used with nGAUSS.m.

```
D U U U 0 0 0 0 0 0          . . . D U U U
L D U U U 0 0 0 0 0          . . L D U U U
L L D U U U 0 0 0 0          . L L D U U U
L L L D U U U 0 0 0          L L L D U U U
0 L L L D U U U 0 0          L L L D U U U
0 0 L L L D U U U 0    ->     L L L D U U U
0 0 0 L L L D U U U          L L L D U U U
0 0 0 0 L L L D U U          L L L D U U .
0 0 0 0 0 L L L D U          L L L D U . .
0 0 0 0 0 0 L L L D          L L L D . . .
```

where

0 = terms outside the bandwidth equal to zero

U = terms in upper triangle within bandwidth

L = terms in lower triangle within bandwidth

D = diagonal terms

. = unused storage area

4.6.2 Specifying Known Values of Y. In program wire.m we diagonalized (or quasi-diagonalized) rows associated with known values of the deflection by multiplying the diagonal by a large number. This ensured that the diagonal was large compared with the off-diagonal terms on the same row. However, in our new code, the diagonals need not be larger than the off-diagonal terms, and may even be small relative to them. Therefore, multiplying them by a large number does not necessarily guarantee that their magnitude will be large. To get around this problem, we obtain a number larger than any off-diagonal term by adding the absolute values of all terms in the matrix. This sum is then multiplied by a suitably large number and used to replace any diagonal associated with a known value of Y.

4.6.3 Nodal D Values. The distributed load specified in the tight wire problem had to be discretized by the user and entered as nodal point loads. The coefficient $D(x)$, which is the equivalent term in our new equation, is numerically integrated, and its discretized values are entered into the right-hand side of the matrix equation. For this reason, the only time a specified nodal value is entered is when the problem specifies a point source term and a node has been placed at that point. This might be the case in some problems, but in most problems it is more likely that the source terms are distributed. This means that for most, if not all, nodes, the specified nodal values for the right-hand side will be zero.

The exception to this occurs at the boundaries. At these points the nodal values represent the flux into or out of the domain shown as q_0 and q_L in Eq. 4.9. Note that if $D(x)$ is nonzero, then the boundary nodes will also have a contribution from that source. One point of confusion for many students is how to specify a zero flux at a boundary. This, of course, is done simply by specifying a zero value for these flux terms.

4.6.4 Shape Functions. Although the shape functions for our new code will be the same as they were for program wire.m, we will use this opportunity to introduce the use of a function statement to define the shape functions. Such functions will be convenient when we consider codes for the analysis of problems described in two dimensions. The particular function that we will use in our new code is

$$\text{function } s = SF(D,\, n, u)$$

The value returned by the function is

$$\frac{d^D N_n(u)}{du^D}$$

Here u = the Gaussian coordinate, n is the shape function number (for node 1 or 2), and D is the degree of the derivative. Note that for $D = 0$, the shape function itself will be calculated. The complete description of the function will be given in a later section.

4.6.5 INCLUDE Codes. The need to have the functions $A(x)$, $B(x)$, $C(x)$, and $D(x)$ available in the code creates certain problems. It would be best not to edit the main code every time there is a change in these functions. One option is to create them as external functions similar to sin(a), log(x), or the above *SF* function. However, the disadvantage of using function statements is that all variables necessary to define the functions must be specified as arguments at the time the function is defined in the main code. That is, they must be anticipated at the time the main code is written. As the complexity of these functions increases, it becomes more likely that there may be a need for a quantity that has not been anticipated.

To avoid these difficulties we use INCLUDE codes. These are codes stored in independent files under an appropriate name. They will be included in the main code wherever the file name appears. Here is an example:

```
% mainA.m
  a = 6;
  b = 3;
  c = (a-b)/(a+b);
  d = sin(c);
  answer = d
```

can be written as

```
% mainB.m
  a = 6;
  b = 3;
  ABC
  answer = d

-----------------

% ABC.m
  c = (a-b)/(a+b);
  d = sin(c);
```

Here, mainB.m and ABC.m are independent programs located in files with these same names and in the same working directory. When mainB.m is run, it simply inserts the contents of file ABC.m at the location ABC, making mainB.m identical to program mainA.m.

It is with this method that the equations used to describe the coefficients in Eq. 4.1 will be written. The file will be named COEF.m. Note that all variables defined in the main code, up to the line where the INCLUDE code is specified, are available for use in the INCLUDE code. Care must be taken, however, not to change any of these variables in COEF.m, as they likely will be used later in the main code.

In codes developed in later chapters, we will have need for other INCLUDE codes. The practice we follow is to name all INCLUDE codes with uppercase letters as we did for COEF.m. Similar structures can be added to codes written in other languages; for example, in FORTRAN, the use of subroutines serves much the same purpose.

4.7 PROGRAM ode2.m

Program ode2.m is the finite element code for solving Eq. 4.1, a second-order linear ordinary differential equation In addition to ode2.m, you will need the following files in your working directory:

Input Data	User's INCLUDE	Supplied Functions
MESH	COEF.m	nGAUSS.m
QUAD		SF.m

MESH contains the user-defined nodal point coordinates and nodal point values used for a specific analysis. The file has exactly the same format as it had for program wire.m. QUAD is a supplied data file that defines the Gaussian points and weights. It is shown in the example problem section following the description of ode2.m. COEF.m is your user INCLUDE code to define the parameters A, B, C, and D. nGAUSS.m is the supplied function that solves the banded, nonsymmetric equations by Gaussian elimination; it is described in Appendix A. SF.m is the supplied function that defines the two shape functions and their derivatives.

The notation in the code for the finite element equations is

$$[SK]\{Y\} = \{Q\}$$

where Q has been used to correspond to the notation used in many formulations for a source term.

Flow Chart

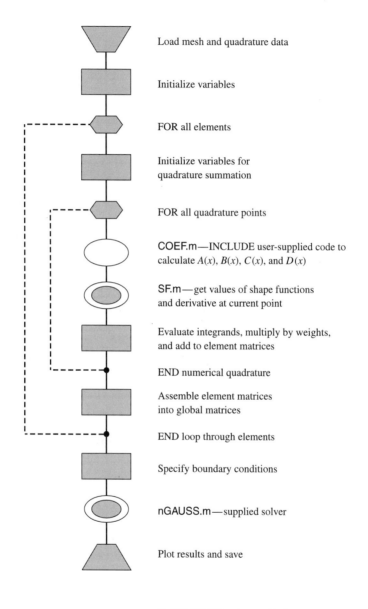

Load mesh and quadrature data

Initialize variables

FOR all elements

Initialize variables for
quadrature summation

FOR all quadrature points

COEF.m—INCLUDE user-supplied code to
calculate $A(x)$, $B(x)$, $C(x)$, and $D(x)$

SF.m—get values of shape functions
and derivative at current point

Evaluate integrands, multiply by weights,
and add to element matrices

END numerical quadrature

Assemble element matrices
into global matrices

END loop through elements

Specify boundary conditions

nGAUSS.m—supplied solver

Plot results and save

= a user-supplied INCLUDE code, outside the current
program, that must be in the current working directory

4.7.2 Code

```
        clear
%       ----------------
%       INPUT DATA
%       ----------------
        load MESH    -ASCII
        load QUAD    -ASCII


%       ----------------
%       Define MESH Data
%       ----------------
        NUMNP = MESH(1,1);
        NUMEL = NUMNP-1;
        for i=1:NUMNP;
            XORD(i) = MESH(i+1,1);
            NPBC(i) = MESH(i+1,2);
            Y(i)    = MESH(i+1,3);
            Q(i)    = MESH(i+1,4);
        end
%       ----------------
%       Define QUAD Data
%       ----------------
        NQPTS = QUAD(1,1);
        for i=1:NQPTS
            GPTS(i)=QUAD(i+1,1);
            GWTS(i)=QUAD(i+1,2);
        end
%       ---------------------------
%       Initialize stiffness matrix
%       ---------------------------
        for I=1:NUMNP
            for J=1:3
                SK(I,J)=0.0;
            end
        end
%       ---------------------------------
%       FORMATION OF STIFFNESS MATRIX
%        and
%       RIGHT HAND SIDE
%       ---------------------------------
        IB=3;
        for I=1:NUMEL
%           ---------------------------
%           Initialize Element Variables
%           ---------------------------
            Xa = XORD(I);
            Xb = XORD(I+1);
            RL=Xb-Xa;
            DxDu=RL/2.0;
            DuDx=2.0/RL;
```

Load mesh data and quadrature data
Both files must be in working directory

NUMNP	Number of nodal points
NUMEL	Number of elements
XORD	x coordinates of nodes
NPBC	Nodal point boundary condition
Y	Nodal point values of $y(x)$
Q	Nodal point right-hand side

NQPTS	Number of quadrature points
GPTS	Gauss points
GWTS	Gauss weights

SK	Global stiffness matrix

IB	Bandwidth of stiffness matrix
I	Current element number

RL	Length of current element
DxDu	$\dfrac{dx}{du}$
DuDx	$\dfrac{du}{dx}$

```
    for J=1:2
      for K=1:2
         S1(J,K)=0.0;
         S2(J,K)=0.0;
         S3(J,K)=0.0;
      end
      Qe(J)=0.0;
    end
%  ------------------------
%  Begin Gaussian Quadrature
%  ------------------------
    for J=1:NQPTS
      u = GPTS(J);
%  --------------------
%  Global coordinate of
%  current Gauss point
%  --------------------
      Xg = (Xa+Xb)/2  + (RL/2.0)*u;

%  ----------------------------
%  INCLUDE COEF.m
%  Defines: AX, BX, CX, and DX
%  ----------------------------
      COEF

%  ----------------------------------
%  Gradient of shape functions at
%  current Gauss point
%  ----------------------------------
      DNDX(1) = SF(1,1,u)*DuDx;
      DNDX(2) = SF(1,2,u)*DuDx;

%  ----------------------------
%  Element stiffness matrices
%  ----------------------------
      Wt = GWTS(J);
      for K=1:2
        for L=1:2
          S1(K,L)=S1(K,L) ...
            +Wt*DNDX(K)*AX*DNDX(L)*DxDu;
          S2(K,L)=S2(K,L) ...
            +Wt*SF(0,K,u)*BX*DNDX(L)*DxDu;
          S3(K,L)=S3(K,L) ...
            +Wt*SF(0,K,u)*CX*SF(0,L,u)*DxDu;
        end
        Qe(K)=Qe(K) + Wt*SF(0,K,u)*DX*DxDu;
      end
    end
%  --------- Quadrature now complete
```

Initialize element stiffness matrices

S1	$[S_1]$
S2	$[S_2]$
S3	$[S_3]$

Qe   Element right-hand side

J   Quadrature point number

u   Quadrature coordinate u

Xg   Global coordinate of current quadrature point

COEF   Include user-written COEF.m code. Defines AX, BX, CX, and DX at current quadrature point.

DNDX(1) $\dfrac{dN_1}{dx} = \dfrac{dN_1}{du}\dfrac{du}{dx}$

DNDX(2) $\dfrac{dN_2}{dx} = \dfrac{dN_2}{du}\dfrac{du}{dx}$

S1(K,L) $\displaystyle\int_0^{L_e} \{N'\}\, A(x)\, \lfloor N' \rfloor\, dx$

S2(K,L) $\displaystyle\int_0^{L_e} \{N\}\, B(x)\, \lfloor N' \rfloor\, dx$

S3(K,L) $\displaystyle\int_0^{L_e} \{N\}\, C(x)\, \lfloor N \rfloor\, dx$

Qe(K) $\displaystyle\int_0^{L_e} \{N\}\, D(x)\, dx$

```
% ----------------------------
% Assemble element matrices
% into global matrix
% ----------------------------
    K1=I-1;
    L0=2;
    for K=1:2
      L0=L0-1;
      K1=K1+1;
      L1=L0;
      for L=1:2
        L1=L1+1;
        SK(K1,L1)=SK(K1,L1) ...
            +S1(K,L)-S2(K,L)-S3(K,L);
      end
      Q(K1) =  Q(K1) + Qe(K);
    end
  end
% ---- Global Matrices are assembled

% ------------------------
% BOUNDARY CONDITIONS
% ------------------------
  B = 0;
  for I=1:NUMNP
    for J=1:3
      B = B+abs(SK(I,J));
    end
  end
  B = B*(1.0E+04);
  for I=1:NUMNP
    if NPBC(I) == 1
      SK(I,2)=B;
      Q(I)=Y(I)*B;
    end
  end
% ------------------------
% CALL EQUATION SOLVER
% ------------------------
  Y =  nGAUSS(SK,Q,NUMNP,IB);

% ----------------------
% OUTPUT DATA
% ----------------------
  save Y Y -ASCII
  plot(XORD,Y)
  grid
  title(' Results from ode.m')
  xlabel(' Independent variable')
  ylabel(' Dependent variable')
```

Compact storage for nonsymmetric, banded matrices

K   Row in element matrix
L   Column in element matrix
K1   Row in global matrix
L1   Column in global matrix

SK   Global stiffness matrix
Q   Global right-hand side

Determine big number compared with any term in *SK* matrix. Use this to replace diagonal term on rows where *Y* value is known.

B $\qquad = \left(\sum \sum \| SK(I,J) \| \right) 10^4$

if NPBC(I) = 1, then
Y value is known at node *I*

SK(I,2)   Stiffness diagonal on row *I*
B   Replace diagonal with B(ig)
Q(I)   Right-hand side for row *I*
Y(I)   Specified value for *Y* obtained from data file MESH

nGAUSS   External equation solver for nonsymmetric banded matrices

Save and plot *Y* values.

4.7.3 Auxiliary Code

In addition to ode2.m, your working directory must have SF.m.

SF.m

```
function s  = SF(D,node,u)
%------------------------------------
%////////////////////////////////////
%------------------------------------
%  Shape Functions
%      D = derivative
%        = 0 function itself
%        = 1 first derivative of function
%  node = nodal point number (1 or 2)
%      u = local coordinate
%------------------------------------

  if D == 0   % Calculate shape function
    if node == 1
      s = (0.5 - 0.5*u);
    elseif node == 2
      s = (0.5 + 0.5*u);
    else
      error('Error #1 in SF function')
    end

  elseif D == 1  % Calculate derivative
    if node == 1
      s = -0.5;
    elseif node == 2
      s = +0.5;
    else
      error('Error #2 in SF function');
    end
  else
    error('Error #3 in SF function')
  end
```

Function that calculates the shape functions, N_1 and N_2, and the first derivative of the shape functions, dN_1/du and dN_2/du, at Gaussian coordinate u

if $D = 0$, $d^0 N/du^0 = N$

$N_1(u) = 0.5 - 0.5u$

$N_2(u) = 0.5 + 0.5u$

if $D = 1$, $d^1 N/du^1 = dN/du$

$dN_1/du = -0.5$

$dN_2/du = +0.5$

4.7.4 Test Problem If a family of finite element approximations contains the exact solution to a problem, then the FEM method will select that approximation; that is, it will give the exact answer. This provides an excellent method to test codes for possible errors. Because piece-wise linear approximations are used in ode2.m, it will give the exact answer to any problem whose solution is a linear function. The following coefficients and data correspond to the solution

$$y = 2x + 1$$

QUAD (Quadrature Data File)

```
%------------------------------------
%   QUAD data file
%
% Number of points     dummy number
%------------------------------------
     3                    0
%------------------------------------
% Coordinates           Weights
%------------------------------------
-0.77459666924148    0.55555555555556
 0.0                 0.88888888888889
+0.77459666924148    0.55555555555556
%------------------------------------
```

Gaussian quadrature data.

The following coordinates and weights are for three-point Gaussian quadrature.

This file can be edited to change the number of points desired. Such changes will not affect ode2.m.

MESH (User-Supplied Data File)

```
%   MESH data file
%-------------------------------
% NUMNP   + 3 dummy numbers
%-------------------------------
     4       0    0    0
%-------------------------------
% XORD   NPBC     Y    Q
%-------------------------------
     1      1     3    0
    1.2     0     0    0
    1.6     0     0    0
     2      1     5    0
%-------------------------------
```

MESH data file. Defines problem.

First line contains only NUMNP. The other numbers are dummy values.

For each node, XORD, NPBC, Y, and Q must be entered.

For NPBC = 1,
    Y must be specified (Q is dummy).

For NPBC = 0,
    Q must be specified (Y is dummy).

COEF.m (User-Supplied INCLUDE Code)

```
%==============================
%    COEF.m  include file
%
%    Coefficients for ode2.m
%
%==============================
        AX  =  Xg^3;
        BX  =  -Xg^2 - Xg;
        CX  =  3*Xg;
        DX  =  -10*Xg^2 - Xg;
```

Sample coefficients to produce the solution
$y(x) = 2x + 1$

$A(x) = x^3$

$B(x) = -x^2 - x$

$C(x) = 3x$

$D(x) = -10x^2 - x$

If ode2.m is run with these files, the following graph will appear:

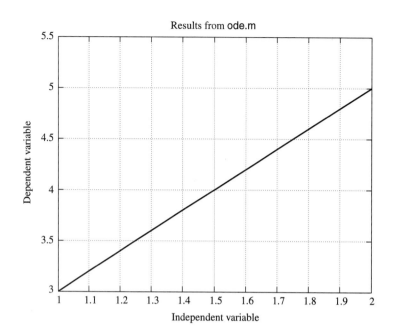

These solutions are exact regardless of the number of elements used. Hence, it is best to make these tests using only a few elements; even two elements will suffice in many cases. The problems are easily defined by first selecting the desired solution and three of the four coefficients, then solving for the fourth coefficient that would be necessary to satisfy the differential equation.

EXERCISES

Study Problems

S1. Write out the complete global stiffness matrix for the wire on the elastic foundation using two elements.

S2. Determine the dimensions of all terms in the final matrix equation for the elastic foundation problem and the pipe flow problem.

S3. In general, the finite element procedure gives only an approximate solution to the exact answer. Therefore, at least one physical law is not being satisfied by the solution. For the wire problem (with or without an elastic foundation), which law is violated? For the pipe flow problem, which law is violated?

S4. In our differential equation, does the term

$$\frac{d}{dx}\left(A(x)\frac{dy}{dx}\right)$$

demand that dy/dx be continuous, or that the function

$$A(x)\frac{dy}{dx}$$

be continuous? Give a physical example to illustrate your answer.

S5. Explain why the integration by parts used to arrive at the weak form of the governing equation places an additional restriction on δy.

S6. The equations shown for the two shape functions in terms of the Gaussian coordinate u are dimensionless. Show that this is true by writing these same equations when the endpoints are $\pm a$ rather than ± 1, where both a and u have dimensions of length.

S7. Carry out all the steps necessary to go from the weak form to the original governing differential equation and its boundary conditions. State all assumptions (restrictions) you have to make in order to do this.

S8. It is almost always the case that for the physical problems we are considering, either the dependent variable, $y(x)$, or the forcing variable, $F(x)$, is known at every point of the system. Thus, one or the other can be specified at every node in a FEM analysis. However, this is not a necessary requirement to obtain a solution. What is necessary, of course, is that there be no more unknowns than there are equations. As a simple example, consider the equation

$$
\begin{bmatrix} 2 & 1 \\ 1 & 2 \end{bmatrix} \begin{Bmatrix} y_1 \\ y_2 \end{Bmatrix} = \begin{Bmatrix} f_1 \\ f_2 \end{Bmatrix}
$$

Is there a solution for the condition $y_1 = 1$ and $f_1 = 3$? If so, what are the values for y_2 and f_2? Is this solution unique? Rewrite the above equation in the form

$$
\begin{bmatrix} a_{11} & a_{12} \\ a_{21} & a_{22} \end{bmatrix} \begin{Bmatrix} f_2 \\ y_2 \end{Bmatrix} = \begin{Bmatrix} y_1 \\ f_1 \end{Bmatrix}
$$

S9. Suppose, as an example of the above, it is necessary to determine the temperature distribution due to pure conduction in a solid rod with a distributed heat source $Q(x)$. At one end, both the temperature and the heat flux are known, but at the other end, neither is known. Is it possible to solve this problem? If there is a solution, could you use ode2.m to obtain it? Consider, for example, a possible modification of ode2.m as well as using ode2.m as it is written, but carrying out an iterative solution.

S10. For problems where $C(x)$ is identically zero, it is necessary to specify at least one value of the dependent variable, y, as a boundary condition. If this is not done, there will not be a unique solution to the problem. Explain this by considering that if $Y(x)$ is a particular solution, then $Y(x) + C$ would also be a solution, even with the prescribed boundary conditions for $A(dy/dx)$ at both ends. Also explain why this is not a problem if $C(x) \neq 0$.

S11. For a finite element analysis, the lack of uniqueness in a solution will be manifested by the stiffness matrix being singular. Examine the three element matrices, $[S_1]$, $[S_2]$, and $[S_3]$, and determine which are and which are not singular. In light of your findings, explain when the requirement for specifying at least one value of y is necessary.

S12. For problems where $C(x) \neq 0$, we have stated that it is not necessary to specify the value of y at either boundary. Such an equation was used to describe the deflection of a tightly stretched wire on an elastic foundation. Specifying y at the ends is the same as specifying that the ends of the wire are supported.

Hence, not specifying y at either end is the same as stating that the ends are not supported, at least not in the vertical direction. If the wire is constrained in the horizontal direction to maintain tension, is it physically possible for it not to have any end support in the vertical direction? Explain.

S13. We have stated that for problems where $C(x) = 0$, there will be no unique solution unless a value of the dependent variable is specified at one of the ends. In many cases this means there will be multiple solutions. However, we also know that there may be no solution. Consider a heat-conducting rod 2 units in length that has a coefficient of thermal conductivity equal to 0.5 units, and a distributed heat source along the rod equal to 2 units, where all units are consistent. Also assume that at each end of the rod $k(d\Theta/dx) = 0$, where Θ is the temperature. Write out the complete fem equation for two elements. Show by the use of Cramer's rule that there is no solution to this problem. Considering that the equation is for steady-state heat conduction, does it make sense that there is no solution rather than multiple solutions? Explain.

Numerical Experiments and Code Development

N1. Write a short MATLAB program that creates a MESH data file using as input the number of elements and the boundary coordinates. The boundary values can be left as zero and the file edited for whatever boundary conditions are desired.

N2. The function

$$y = 1 + x$$

satisfies the equation

$$\frac{d^2y}{dx^2} + y = (1 + x)$$

Assume the range of analysis is $0 \le x \le 6$, and determine the appropriate boundary values for y and $A(dy/dx)$. Use a mesh of six elements of the same length and run ode2.m for the following boundary conditions:

$y(x = 0)$	$y(x = L)$
Known	Known
Known	Unknown
Unknown	Known
Unknown	Unknown

All solutions should agree with the exact answer.

N3. The function

$$y = 1 + x$$

satisfies the equation

$$\frac{d^2y}{dx^2} + \frac{dy}{dx} = 1$$

Assume the range of analysis is $0 \le x \le 6$, and determine the appropriate boundary values for y and $A(dy/dx)$. Use a mesh of six elements of the same length and run ode2.m for the following boundary conditions:

$y(x = 0)$	$y(x = L)$
Known	Known
Known	Unknown
Unknown	Known
Unknown	Unknown

Discuss the results for all four analyses. Explain how and why the boundary condition specification affected your solution.

N4. The function

$$y = 1 + x^2$$

satisfies the equation

$$A_c \frac{d^2y}{dx^2} + \frac{dy}{dx} = 2(A_c + x)$$

for any value of the constant A_c. Assume the range of analysis is $0 \le x \le 4$, and use the appropriate boundary conditions for y known at $x = 0$ and $A_c(dy/dx)$ known at $x = 4$ and set $A_c = 0.50$. Run ode2.m using 50, 100, and 500 elements. For each case, plot percent error as a function of x. Run the same problem using 50 elements and specify $y(x = L)$ also known.

N5. Given the equation

$$\frac{d^2y}{dx^2} - \frac{dy}{dx} = 0$$

with boundary conditions

$$y(x = 0) = 50 \quad \text{and} \quad y(x = L) = 10$$

use ode2.m with 20 elements to solve $y(x)$ for $L = 10$, 100, and 1000. This problem illustrates the effect of element length on oscillations of the solution.

N6. Revise program wire.m to accommodate an elastic foundation for a constant foundation modulus. Run your new program for $T = 20{,}000$ lb, $k = 2000$ lb/ft^2, and $L = 20$ ft. Place a concentrated force of 1 lb at the center of the wire. Compare your results with the deflection without a foundation.

N7. The matrices developed for constant coefficients can be used for variable coefficients where the value of the coefficient is taken at the center of the element. Revise program wire.m with this technique to solve

$$\frac{d}{dx}\left[A(x)\frac{dy}{dx}\right] + C(x)y + D(x) = 0.0$$

Note that $B = 0$ is assumed here so the resulting matrix will be symmetric. Consider several ordinary differential equations of your own choosing, and compare solutions found with your new code with those found using ode2.m. Is the approach described above equivalent to one-point Gaussian quadrature?

N8. Revise ode2.m to include one or more of the following:
   (a) After solving for the nodal point values, search through them to determine maximum and minimum values. Include as output these values and their locations.
   (b) Calculate the values of $A(dy/dx)$ at the center of each element and the x coordinates at these points and plot your results.
   (c) Include in ode2.m a routine for calculating the external fluxes, i.e., the Q values, at all points where they were not specified.

Project

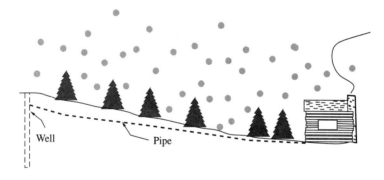

P1. It is necessary to prevent water in a pipe from freezing during the winter months. The pipe runs from an artesian well to a cabin 2.17 km down a hillside. The insulation for the pipe consists partly of ground cover and partly of some commercial wrapping material. A fair approximation to the insulation can be given by

$$h = h_o\left(1.3 - \frac{x}{L}\right) \tag{4.58}$$

where h is the coefficient of convective heat transfer, L is the total length of the pipe, and x is the distance from the well. The temperature in the pipe can be described by the equation

$$\frac{d}{dx}\left(k\frac{d\Theta}{dx}\right) - v(\rho C_p)\frac{d\Theta}{dx} + h(\Theta_a - \Theta) = 0.0 \tag{4.59}$$

where

    Θ is the temperature (°C)

    Θ_a is the ambient temperature $= -30\,°C$.

    v is the velocity in the pipe (m/s)

    ρ is the density $= 1.0E+03\ \text{kg/m}^3$

    C_p is the specific heat $= 4.211E+03\ \text{N m/kg}\,°C$

    k is the thermal conductivity $= 0.574\ \text{N m/s m}\,°C$

    h is the coefficient of convective heat transfer, $\text{N m/m}^3\ \text{s}\,°C$

In order to keep the water from freezing, it is necessary to maintain a small flow in the pipe throughout the winter months. What is the minimum average velocity that will keep the water from freezing before it reaches the cabin? The water enters the pipe from the well at a temperature of 20°C. It empties into a storage tank in the cabin, which contains water at approximately the temperature of the entering water. Assume, therefore, that $k(d\Theta/dx) = 0$ at that point. Take $h_o = 4.5\ \text{N m/m}^3\ \text{s}\,°C$.

A FINITE ELEMENT FUNCTION FOR TWO DIMENSIONS

5.1 BASIC IDEA

In this chapter we introduce a finite element function to approximate surfaces in two dimensions. In the following chapter, we will use these functions to approximate surfaces defined by partial differential equations.

Consider the two-dimensional extension of the piecewise linear function we used for our one-dimensional problems. Whereas in one dimension we approximated a curve with a series of straight-line segments, we will now approximate a surface with a system of triangular flat plates. Figure 5.1 illustrates the concept we are developing.

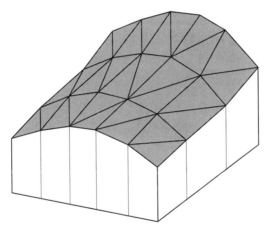

Figure 5.1. Finite element approximation of a surface.

It is important to note that the piecewise approximation shown in Fig. 5.1 is completely defined by the nodal point values of its 24 nodes. That is, it would be possible to construct such a surface by having knowledge of only the nodal point coordinates and surface "elevations" at each node. A variety of sufaces could be constructed simply by changing the elevations at each node. Clearly, as the number of nodal points increased, it would become possible to model ever more accurately any given surface.

Each segment of the surface is a plane. The projection of each of these segments onto the x-y plane is a triangle within which the surface varies linearly with x and y. Because the segments are planes, their edges are straight. Therefore, two adjacent segments line up perfectly along their common boundary. This ensures that a surface thus defined is continuous everywhere. As we consider how to use such a surface to approximate a solution to a partial differential equation, we will see this is an important characteristic.

5.2 MATHEMATICAL DESCRIPTION

The above discussion gave us a physical feel for finite element approximations in two dimensions; we must now consider how to describe such approximations mathematically. We begin with an early procedure which is, mathematically, straight forward. Following that, we introduce a second method that will be useful when we consider elements other than triangles.

5.2.1 Classical Approach. Figure 5.2 represents a typical element with a local coordinate system located at one of its nodes. Let Φ be the dependent variable to be approximated and let us write

$$\Phi(x, y) = a_0 + a_1 x + a_2 y \tag{5.1}$$

or, in matrix notation

$$\Phi(x, y) = \lfloor 1 \quad x \quad y \rfloor \begin{Bmatrix} a_0 \\ a_1 \\ a_2 \end{Bmatrix} = \lfloor M \rfloor \{a\} \tag{5.2}$$

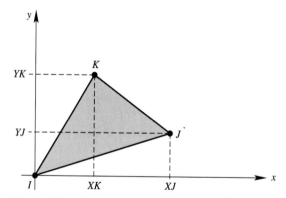

Figure 5.2. Typical element with local coordinates.

Here the a's are the undetermined parameters of our approximating function. As we did for our one-dimensional approximation, we now want to define our function in terms of its nodal values rather than in terms of the a values. We determine the relationship between the a's and the three nodal values of Φ using Eq. 5.2 and the nodal coordinates. Thus,

$$\begin{Bmatrix} \Phi_I \\ \Phi_J \\ \Phi_K \end{Bmatrix} = \begin{bmatrix} 1 & 0 & 0 \\ 1 & XJ & YJ \\ 1 & XK & YK \end{bmatrix} \begin{Bmatrix} a_0 \\ a_1 \\ a_2 \end{Bmatrix} \tag{5.3}$$

which we abbreviate as

$$\{\Phi\}_e = [A]\{a\} \tag{5.4}$$

Inversion of $[A]$ gives us

$$\{a\} = [A]^{-1}\{\Phi\}_e \tag{5.5}$$

and

$$\Phi(x, y) = \lfloor M \rfloor [A]^{-1} \{\Phi\}_e \tag{5.6}$$

where

$$[A]^{-1} = \frac{1}{D}\begin{bmatrix} D & 0 & 0 \\ (YJ - YK) & YK & -YJ \\ (XK - XJ) & -XK & XJ \end{bmatrix} \tag{5.7}$$

and

$$D = \|A\| = (XJ \times YK - XK \times YJ) = 2 \times \text{AREA} \tag{5.8}$$

We now can write

$$\lfloor N \rfloor = \lfloor M \rfloor [A]^{-1} \tag{5.9}$$

and

$$\Phi(x, y) = \lfloor N_I(x, y) \quad N_J(x, y) \quad N_K(x, y) \rfloor \begin{Bmatrix} \Phi_I \\ \Phi_J \\ \Phi_K \end{Bmatrix} \tag{5.10}$$

where N_I, N_J, N_K are the shape functions. Figure 5.3 illustrates N_J. Shape functions N_I and N_K have identical shapes. The subscripts I, J, and K are often replaced with the subscripts 1, 2, and 3; both notations are used in this text.

5.2.2 A Second Approach. We now consider a second approach for obtaining the shape functions. First, we recognize three characteristics that each shape function must have:

1. Each must be a linear function of x and y.
2. Each must be equal to unity at the node it represents.
3. Each must be equal to zero along the opposite side.

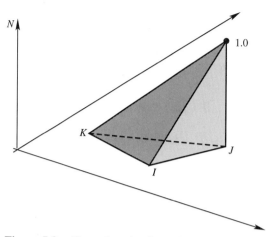

Figure 5.3. Shape function for node J.

The first two characteristics are somewhat obvious. The third is a bit more subtle. In order to ensure continuity of our approximating function between adjacent elements, it is necessary that only the nodes contiguous to a side determine the function along that side. This is necessary to ensure that two elements that have a common boundary also share the only nodal values used to approximate the function along their common boundary.

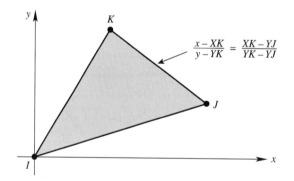

Figure 5.4. Equation for the line *JK*.

With this information, we now obtain the shape function for node *I*. Consider Fig. 5.4. The equation for the line connecting points *J* and *K* of the opposite side is

$$\frac{x - XK}{y - YK} = \frac{XK - XJ}{YK - YJ} \tag{5.11}$$

or

$$(x - XK)(YK - YJ) - (y - YK)(XK - XJ) = 0 \tag{5.12}$$

If we use this function as our shape function, then it will be linear in x and y and zero along side JK. The only requirement not met is that it be unity at node I. This is easily taken care of by scaling the function. Hence, we let

$$N_I(x, y) = C[(x - XK)(YK - YJ) - (y - YK)(XK - XJ)] \tag{5.13}$$

and select a value for C that makes the function equal to unity at node I, that is, at $(x, y) = (0, 0)$. This gives us

$$C = \frac{1}{XK(YJ) - YK(XJ)} = -\frac{1}{D} \tag{5.14}$$

Here the denominator turns out to be the determinate of $[A]$ as was also given in Eq. 5.8. Its value is equal to twice the area of the element. We now have for our final equation

$$N_I = -\frac{1}{D}[(x - XK)(YK - YJ) - (y - YK)(XK - XJ)] \tag{5.15}$$

which has a shape similar to that illustrated in Fig. 5.3. The equations for the two remaining shape functions can be obtained in the same way. All three functions are given in the summary of this chapter.

5.3 ADDITIONAL RELATIONSHIPS

Before we are able to use this piecewise function to obtain an approximate solution to a partial differential equation, we need to develop some additional relationships.

5.3.1 The Gradient of Φ Most of the weak forms of the partial differential equations that we will be considering contain the gradient of the dependent variable; that is,

$$\{\nabla\Phi\} = \left\{ \begin{array}{c} \dfrac{\partial\Phi}{\partial x} \\[2ex] \dfrac{\partial\Phi}{\partial y} \end{array} \right\} \tag{5.16}$$

Within a single element, the above terms have the form

$$\frac{\partial\Phi}{\partial x} = \left| \begin{array}{ccc} \dfrac{\partial N_I}{\partial x} & \dfrac{\partial N_J}{\partial x} & \dfrac{\partial N_K}{\partial x} \end{array} \right| \left\{ \begin{array}{c} \Phi_I \\[1ex] \Phi_J \\[1ex] \Phi_K \end{array} \right\} \tag{5.17}$$

$$\frac{\partial \Phi}{\partial y} = \left\lfloor \frac{\partial N_I}{\partial y} \quad \frac{\partial N_J}{\partial y} \quad \frac{\partial N_K}{\partial y} \right\rfloor \left\{ \begin{array}{c} \Phi_I \\ \Phi_J \\ \Phi_K \end{array} \right\} \tag{5.18}$$

These can be combined into one matrix equation as follows:

$$\left\{ \nabla \Phi \right\} = \left\{ \begin{array}{c} \dfrac{\partial \Phi}{\partial x} \\[2mm] \dfrac{\partial \Phi}{\partial y} \end{array} \right\} = \left[\begin{array}{ccc} \dfrac{\partial N_I}{\partial x} & \dfrac{\partial N_J}{\partial x} & \dfrac{\partial N_K}{\partial x} \\[2mm] \dfrac{\partial N_I}{\partial y} & \dfrac{\partial N_J}{\partial y} & \dfrac{\partial N_K}{\partial y} \end{array} \right] \left\{ \begin{array}{c} \Phi_I \\ \Phi_J \\ \Phi_K \end{array} \right\} \tag{5.19}$$

which we abbreviate

$$\left\{ \nabla \Phi \right\} = [N']\left\{ \Phi \right\} \tag{5.20}$$

5.3.2 The Test Function. We will also use our finite element functions to represent the test, or weighting, function used in the weak form of our governing equations. As in the last chapter, we will usually interpret this function as an arbitrary variation of the trial, or approximating, function. Thus, we have

$$\delta \Phi(x, y) = \lfloor N \rfloor \left\{ \delta \Phi \right\} \tag{5.21}$$

and

$$\left\{ \delta \nabla \Phi \right\} = [N']\left\{ \delta \Phi \right\} \tag{5.22}$$

5.4 SUMMARY

$$N_I = -\frac{1}{D}[(x - XK)(YK - YJ) - (y - YK)(XK - XJ)]$$

$$N_J = +\frac{1}{D}[x(YK) - y(XK)] \tag{5.23}$$

$$N_K = -\frac{1}{D}[x(YJ) - y(XJ)]$$

$$\frac{dN_I}{dx} = -\frac{YK - YJ}{D} \qquad \frac{dN_I}{dy} = +\frac{XK - XJ}{D}$$

$$\frac{dN_J}{dx} = +\frac{YK}{D} \qquad \frac{dN_J}{dy} = -\frac{XK}{D} \qquad (5.24)$$

$$\frac{dN_K}{dx} = -\frac{YJ}{D} \qquad \frac{dN_K}{dy} = +\frac{XJ}{D}$$

$$D = (XJ)(YK) - (XK)(YJ)$$

EXERCISES

Study Problems

S1. Determine the inverse of $[A]$.

S2. Show that the determinate of $[A]$ is equal to twice the area of the triangular element.

S3. Derive the shape functions for nodal points J and K directly as was done for node I.

S4. Determine explicit formulas for the six terms in $[N']$.

S5. Using Fig. 5.3 as a reference, make a freehand sketch of the following surfaces:
   (a) $N_K + N_J$
   (b) $N_K + N_J + N_I$
   (c) $2N_K + N_J$

S6. Show that $D = -2(\text{AREA})$ if the nodes are labeled IJK in a clockwise direction.

S7. Sum the equations for N_I, N_J, and N_K and show that the sum is identically equal to 1.0.

S8. For

XJ	YJ	XK	YK
12	8	6	16

Calculate N_I, N_J, and N_K for

Point	x	y
1	6	4
2	6	8

Make a scale drawing of the element and show the two points given above. Comment on the values you found for N_I, N_J, and N_K. Are they what you would expect? Explain.

POISSON'S EQUATION: FEM APPROXIMATION

In this chapter we use the two-dimensional finite element function to obtain approximate solutions to boundary value problems described by Poisson's and Laplace's equations. These equations are adequate to describe a large number of applied problems in engineering, some of which are described in the next chapter.

6.1 GOVERNING EQUATIONS

We begin with a description of the partial differential equation (or strong form) and then develop the weak form suitable for finite element analysis.

6.1.1 Strong Form. The following equation and boundary conditions (along with Fig. 6.1) define the two-dimensional problem we wish to solve by the finite element method.

$$\frac{\partial}{\partial x}\left[k_x\frac{\partial \Phi}{\partial x}\right] + \frac{\partial}{\partial y}\left[k_y\frac{\partial \Phi}{\partial y}\right] = -Q(x, y) \quad \text{in } V \tag{6.1}$$

$$\left.\begin{aligned} k_x\frac{\partial \Phi}{\partial x}n_x + k_y\frac{\partial \Phi}{\partial y}n_y &= q^* \\ \Phi &= \Phi^* \end{aligned}\right\} \text{ on } S \tag{6.2}$$

where

n_x, n_y are the x and y components of the outward unit normal vector to S

q^*, Φ^* are the boundary values of the quantities shown

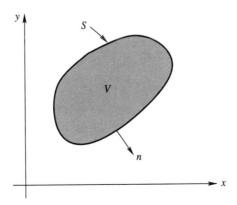

Figure 6.1. Domain of analysis.

A comment about notation: We have chosen to use V (for volume) and S (for surface) to indicate the domain of analysis and its boundary. Clearly, for a two-dimensional analysis, A (for area) might appear more appropriate. However, for most of the applied problems that we consider, the domain of analysis will be a volume of unit thickness. Hence, in order to emphasize this dimensionality, we use the symbols V and S.

A comment about boundary conditions: Equation 6.2 defines two quantities associated with the boundary. For the problems we consider in the next chapter, one or the other, but not both, will be specified at all points on the boundary. In addition, for the problem to be well defined, Φ^* must be specified at at least one point on the boundary. It is customary to specify boundaries according to which one of the two conditions is specified: S_Φ for boundaries where Φ^* is specified, and S_q for boundaries where q^* is specified. Keep in mind that it is not a question of which of the two quantities is known. There are some problems (although it is not usually the case) where the analyst might know the value of both quantities at a given point.[1] It is a question of which quantity is specified in the mathematical formulation of the particular problem under investigation. At this time, however, it is nether necessary nor desirable to make that distinction.

6.1.2 Weak Form. To obtain the weak form, note that

$$\int_V \delta\Phi \left\{ \frac{\partial}{\partial x}\left[k_x \frac{\partial \Phi}{\partial x}\right] + \frac{\partial}{\partial y}\left[k_y \frac{\partial \Phi}{\partial y}\right] + Q(x, y) \right\} dV = 0 \tag{6.3}$$

must be true for any $\delta\Phi$. Integration of this equation by parts gives

$$\int_V \left\{ \frac{\partial}{\partial x}\left[\delta\Phi k_x \frac{\partial \Phi}{\partial x}\right] + \frac{\partial}{\partial y}\left[\delta\Phi k_y \frac{\partial \Phi}{\partial y}\right] + -\left(\frac{\partial \delta\Phi}{\partial x} k_x \frac{\partial \Phi}{\partial x} + \frac{\partial \delta\Phi}{\partial y} k_y \frac{\partial \Phi}{\partial y}\right) + \delta\Phi Q(x, y) \right\} dV = 0 \tag{6.4}$$

Application of Green's theorem,

$$\int_V \left[\frac{\partial P}{\partial x} + \frac{\partial Q}{\partial y}\right] dV = \int_S \left[P\, dy - Q\, dx\right] \tag{6.5}$$

to the first two terms in Eq. 6.5 gives

$$\int_S \left[\delta\Phi k_x \frac{\partial \Phi}{\partial x} n_x + \delta\Phi k_y \frac{\partial \Phi}{\partial y} n_y\right] dS - \int_V \left[\frac{\partial \delta\Phi}{\partial x} k_x \frac{\partial \Phi}{\partial x} + \frac{\partial \delta\Phi}{\partial y} k_y \frac{\partial \Phi}{\partial y} - \delta\Phi Q\right] dV = 0.0 \tag{6.6}$$

From the boundary condition expressed in Eq. 6.2, we can write

$$\int_V \left[\frac{\partial \delta\Phi}{\partial x} k_x \frac{\partial \Phi}{\partial x} + \frac{\partial \delta\Phi}{\partial y} k_y \frac{\partial \Phi}{\partial y} - \delta\Phi Q\right] dV - \int_S \delta\Phi q^* \, dS = 0.0 \tag{6.7}$$

or, in the more usual form,

$$\int_V \left[\frac{\partial \delta\Phi}{\partial x} k_x \frac{\partial \Phi}{\partial x} + \frac{\partial \delta\Phi}{\partial y} k_y \frac{\partial \Phi}{\partial y}\right] dV = \int_V \delta\Phi Q \, dV + \int_S \delta\Phi q^* \, dS \tag{6.8}$$

It is this form of the equation that we will use to obtain our finite element approximations.

---

[1]Certainly, after an analysis, the analyst will know both quantities at all points.

Before proceeding, we show that Eq. 6.8 can be used to obtain the corresponding functional associated with Eqs. 6.1 and 6.2. To do so, we note that

$$\frac{\partial \delta \Phi}{\partial x} k_x \frac{\partial \Phi}{\partial x} + \frac{\partial \delta \Phi}{\partial y} k_y \frac{\partial \Phi}{\partial y} = \delta \frac{1}{2} \left[k_x \left(\frac{\partial \Phi}{\partial x} \right)^2 + k_y \left(\frac{\partial \Phi}{\partial y} \right)^2 \right] \tag{6.9}$$

and

$$\delta(\Phi)Q = \delta(\Phi Q)$$

$$\delta(\Phi)q^* = \delta(\Phi q^*) \tag{6.10}$$

Hence, we can write

$$\delta \left\{ \int_V \left[\frac{1}{2} k_x \left(\frac{\partial \Phi}{\partial x} \right)^2 + \frac{1}{2} k_y \left(\frac{\partial \Phi}{\partial y} \right)^2 - \Phi Q \right] dV - \int_S \Phi q^* \, dS \right\} = 0 \tag{6.11}$$

which is the variational principle associated with Poisson's equation. We will, on occasion, use the symbol J for this functional and state Eq. 6.8 as

$$\delta J = 0 \tag{6.12}$$

6.2 FINITE ELEMENT APPROXIMATION

We now consider the finite element approximation of the solution of our governing equation. We begin by placing Eq. 6.8 in the following matrix form:

$$\int_V \left\lfloor \delta \frac{\partial \Phi}{\partial x} \quad \delta \frac{\partial \Phi}{\partial y} \right\rfloor \begin{bmatrix} k_x & 0 \\ 0 & k_y \end{bmatrix} \begin{Bmatrix} \frac{\partial \Phi}{\partial x} \\ \frac{\partial \Phi}{\partial y} \end{Bmatrix} dV - \int_V \delta \Phi Q \, dV - \int_S \delta \Phi q^* \, dS = 0 \tag{6.13}$$

or, in abbreviated notation,

$$\int_V \lfloor \delta \nabla \Phi \rfloor [R] \{ \nabla \Phi \} \, dV - \int_V \delta \Phi Q \, dV - \int_S \delta \Phi q \, dS = 0 \tag{6.14}$$

where

$$\{ \nabla \Phi \} = \begin{Bmatrix} \dfrac{\partial \Phi}{\partial x} \\ \dfrac{\partial \Phi}{\partial y} \end{Bmatrix} \tag{6.15}$$

Recall from Chapter 5 that within a single element,

$$\Phi(x, y) = \lfloor N \rfloor \{\Phi\}$$

(6.16)

and that

$$\{\nabla\Phi\} = \begin{Bmatrix} \dfrac{\partial\Phi}{\partial x} \\[2mm] \dfrac{\partial\Phi}{\partial y} \end{Bmatrix} = \begin{bmatrix} \dfrac{\partial N_I}{\partial x} & \dfrac{\partial N_J}{\partial x} & \dfrac{\partial N_K}{\partial x} \\[3mm] \dfrac{\partial N_I}{\partial y} & \dfrac{\partial N_J}{\partial y} & \dfrac{\partial N_K}{\partial y} \end{bmatrix} \begin{Bmatrix} \Phi_I \\[2mm] \Phi_J \\[2mm] \Phi_K \end{Bmatrix}$$

(6.17)

abbreviated as

$$\{\nabla\Phi\} = [N']\{\Phi\}$$

(6.18)

Also recall that

$$\delta\Phi = \lfloor N \rfloor \{\delta\Phi\} \quad \text{and} \quad \{\delta\nabla\Phi\} = [N']\{\delta\Phi\}$$

We now substitute the above FEM approximations into Eq. 6.14 and integrate. We must do so, of course, element by element. We begin with the volume integral associated with the stiffness term, where, for our generic element e, we have

$$\int_{V_e} \lfloor \delta\nabla\Phi \rfloor [R] \{\nabla\Phi\} \, dV = \int_{V_e} \lfloor \delta\Phi \rfloor [N']^T [R] [N'] \{\Phi\} \, dV$$

(6.19)

$$= \lfloor \delta\Phi \rfloor [K]_e \{\Phi\}$$

In expanded notation the element stiffness matrix is

$$[K]_e = \int_{V_e} \begin{bmatrix} \dfrac{\partial N_I}{\partial x} & \dfrac{\partial N_I}{\partial y} \\[3mm] \dfrac{\partial N_J}{\partial x} & \dfrac{\partial N_J}{\partial y} \\[3mm] \dfrac{\partial N_K}{\partial x} & \dfrac{\partial N_K}{\partial y} \end{bmatrix} \begin{bmatrix} k_x & 0 \\[3mm] 0 & k_y \end{bmatrix} \begin{bmatrix} \dfrac{\partial N_I}{\partial x} & \dfrac{\partial N_J}{\partial x} & \dfrac{\partial N_K}{\partial x} \\[3mm] \dfrac{\partial N_I}{\partial y} & \dfrac{\partial N_J}{\partial y} & \dfrac{\partial N_K}{\partial y} \end{bmatrix} dV$$

(6.20)

a 3×3 matrix. The other volume integral is

$$\int_{V_e} \delta\Phi Q \, dV = \int_{V_e} \lfloor \delta\Phi \rfloor \lfloor N \rfloor^T Q(x, y) \, dV = \lfloor \delta\Phi \rfloor \{Q\}_e$$

(6.21)

where

$$\{Q\}_e = \int_{V_e} \begin{Bmatrix} N_I \\ N_J \\ N_K \end{Bmatrix} Q(x, y)\, dV \tag{6.22}$$

The total contribution from element e to the volume integration in Eq. 6.8 is

$$\lfloor \delta\Phi \rfloor [K]_e \{\Phi\} - \lfloor \delta\Phi \rfloor \{Q\}_e \tag{6.23}$$

where the Φ terms are those associated with the given element.

We now evaluate the surface integral in Eq. 6.14. Figure 6.2 illustrates a finite element approximation of Φ along a boundary. Consider a typical segment of S, S_s, between nodal point m and nodal point n. Along m-n, Φ is a linear function of S. We may therefore use the shape functions developed for our one-dimensional problems and write

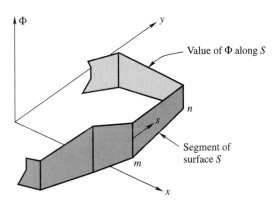

Figure 6.2. Segment of boundary.

$$\Phi(S) = \left\lfloor \left(1 - \frac{S}{\ell_s}\right) \left(\frac{S}{\ell_s}\right) \right\rfloor \begin{Bmatrix} \Phi_m \\ \Phi_n \end{Bmatrix} = \lfloor N_s \rfloor \{\Phi\}_e \tag{6.24}$$

where ℓ_s is the length of the segment.

The surface integral along S_s can now be written as

$$\int_{S_s} q^* \delta\Phi\, dS = \int_{S_s} \lfloor \delta\Phi \rfloor \{N_s\} q(S)\, dS \tag{6.25}$$

which gives us

$$\int_{S_s} q^* \, \delta\Phi \, dS = \lfloor \delta\Phi \rfloor \{Q\}_s \tag{6.26}$$

where

$$\{Q\}_s = \begin{Bmatrix} Q_m \\ Q_n \end{Bmatrix} = \int_{S_s} \begin{Bmatrix} \left(1 - \dfrac{S}{\ell_s}\right) \\ \left(\dfrac{S}{\ell_s}\right) \end{Bmatrix} q(S) \, dS \tag{6.27}$$

The geometrical interpretation for Q_m and Q_n is that they equal the first moment of area of $q(S)$ about their respective nodal point, divided by the length of the segment. Hence, we place a point Q at each node that gives the same moment as the distributed q. Also, note that

$$Q_m + Q_n = \int_{S_s} q \, dS \tag{6.28}$$

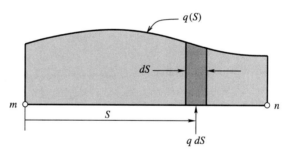

Figure 6.3.   Integration of surface integral

We now have the contributions to Eq. 6.8 from a typical volume element e and a typical surface segment s. We write these as

$$\delta J_e = \lfloor \delta\Phi \rfloor [K]_e \{\Phi\} + \lfloor \delta\Phi \rfloor \{Q\}_e \tag{6.29}$$

and

$$\delta J_s = -\lfloor \delta\Phi \rfloor \{Q\}_s \tag{6.30}$$

To obtain the total variation, we need to sum all such contributions; hence,

$$\delta J = \sum \delta J_e + \sum \delta J_s \tag{6.31}$$

$$\delta J = \lfloor \delta\Phi \rfloor [K]_e \{\Phi\} - \lfloor \delta\Phi \rfloor \{Q\} \tag{6.32}$$

where $[K]$ is the matrix sum of all the $[K]_e$, and $\{Q\}$ is the sum of all the $\{Q\}_e$ and $\{Q\}_s$. We now seek values of $\{\Phi\}$ for which $\delta J = 0$ for all $\{\delta\Phi\}$. This requires

$$[K]_e \{\Phi\} = \{Q\} \tag{6.33}$$

which is our final governing finite element equation.

6.3 PROGRAMMING PRELIMINARIES

Before we look at the program to create and solve Eq. 6.33, several programming details must be discussed. Other details will be given with the listing of the code, and yet others you will be able to obtain from the example data and functions included with the test problem.

6.3.1 Element Identification. As in the codes for one-dimensional analyses, the core of our new code will be the integration of the stiffness matrix, element by element. To accomplish this, we must identify the location and geometry of each element. This is done by identifying the three nodes associated with each element and their coordinates. As an example, consider the finite element mesh shown in Fig. 6.4, where the element numbers are circled. The nodes associated with each of these elements can be identified using the following two-dimensional array:

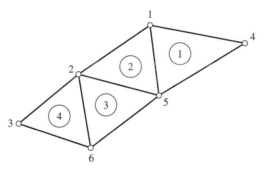

Figure 6.4. A simple two-dimensional mesh.

I	J	K	
1	5	4	Element 1
1	2	5	Element 2
6	5	2	Element 3
2	3	6	Element 4

The nodal point coordinates will be located in two arrays, XORD and YORD.

6.3.2 Integration of $[K]_e$ and $\{Q\}_e$ matrices. Because our shape functions $\lfloor N \rfloor$ are linear in x and y, the gradient terms in $[N']$ are constants. If we also assume that the coefficients k_x and k_y, in $[R]$ are constant, we can write

$$[K]_e = \int_{V_e} [N']^T [R] [N'] dV$$

$$= [N']^T [R] [N'] \int_{V_e} dV \qquad (6.34)$$

$$= [N']^T [R] [N'] V$$

Note again the notation we have chosen to use for our region of integration. In the above, V is actually the area of the element. The notation indicates that we are assuming it is an area associated with a volume of unit thickness.

The derivatives of the shape functions will be stored in the arrays DNDX and DNDY, where, for example;

$$DNDX(2) = \frac{\partial N_2}{\partial x}$$

The element stiffness matrix, $[K]$, will be stored in the 3×3 array S(J,K), defined in terms of the DNDX and DNDY arrays as

$$S(J,K) = (DNDX(J)*RX*DNDX(K)$$
$$\qquad\qquad (6.35)$$
$$+ DNDY(J)*RY*DNDY(K))*VOL$$

where

$$VOL = ((XJ*YK) - (XK*YJ))/2.0 \qquad (6.36)$$

The expression for $\{Q\}_e$ is

$$\{Q\}_e = \int_{V_e} \{N \} Q \, dV \qquad (6.37)$$

If we assume Q constant within each element, it can be shown that all three components of $\{Q\}_e$ are equal and have the value

$$\{Q \}_e = \left\{ \begin{array}{c} Q*VOL/3.0 \\ Q*VOL/3.0 \\ Q*VOL/3.0 \end{array} \right\} \qquad (6.38)$$

6.3.3 The Global Stiffness Matrix. Placement of an element S matrix into the global SK matrix is best explained by an example. Let

$$[S]_e = \begin{bmatrix} S_eII & S_eIJ & S_eIK \\ S_eJI & S_eJJ & S_eJK \\ S_eKI & S_eKJ & S_eKK \end{bmatrix} \tag{6.39}$$

be a typical stiffness matrix. If, say, it is for element 3 shown in Fig. 6.4, the nodes $\{I, J, K\}$ correspond to the nodal numbers $\{6, 5, 2\}$ as obtained from the NP array. Thus, the nine terms of this matrix would appear in the global stiffness matrix as:

	1	2	3	4	5	6
1		S_3KK			S_3KJ	S_3KI
2						
3						
4						
5		S_3JK			S_3JJ	S_3JI
6		S_3IK			S_3IJ	S_3II

The matrix is symmetric and will also be banded; thus, appropriate storage will be used.

6.3.4 User-Provided INCLUDE Codes. As our finite element codes become more complex, it is necessary to use more user-written INCLUDE codes to define specific problems. In the code to follow, poisson.m, there are two such INCLUDE codes: INITIAL.m and COEF.m.

INITIAL.m. In codes wire.m and ode2.m, the initialization of variables is made by loading the MESH data file. As we now consider two-dimensional problems, it becomes more difficult to define the initial values of the variables this way. It is much more convenient to initialize variables using an INCLUDE code. This will appear in poisson.m after the mesh data has been loaded and after the general initialization has taken place. In INITIAL.m the user must define NPBC, PHI, and Q for all nodes that differ from the values given in the general initialization (all zeros). This will be accomplished through the use of a nodal point code defined in the MESH data file exactly as the NPBC array was defined in wire.m and ode.m. However, in poisson.m the NPcode array, as it will be called, is not used. Only in INITIAL.m will it be used to identify which nodes should have which boundary conditions. The NPcode values are assigned by the user and entered with the mesh data. These values can then be used in the user's own INITIAL.m code to assign specific boundary conditions. An example of both a data file and an INITIAL.m code will be given with the test problem.

COEF.m. The three parameters, $k_x(x, y)$, $k_y(x, y)$, and $Q(x, y)$, are defined for each element with the COEF.m INCLUDE. Available for the definitions will be the coordinates of the centroid of the element, XC and YC, as well as the element number, the NP array, etc. The parameters are assumed constant within the element for the integrations. This provides a good approximation for the integration of the stiffness matrix where all other terms in the integrand are constant. However, it is not as accurate a procedure for integration of the right-hand side. Care must be taken to make elements small enough that $Q(x, y)$ is approximately constant within each element, or at least approximately linear.

6.3.5 The MESHo, NODES, and NP Data Files. Three files define the mesh used for the analysis. The first, MESHo, contains three separate lines:

NUMNP = Number of nodal points

NUMEL = Number of elements

NNPE = Number of nodes per element

The number of nodes per element for poisson.m is 3. For codes to be considered later, higher-order elements will be used for which the number of nodes per element will be greater than 3.

The second file, NODES, contains one line per node. Each line gives for its particular node

XORD   YORD   NPcode

that is, the x and y coordinates of the node and the node identification number.

The third file, NP, contains the connectivity array for each element explained earlier. An example of such an array was given in Section 6.3.1.

One of the problems encountered in using finite element codes for two-dimensional analyses is how to create the mesh data. It is not unusual for there to be several thousand nodes and elements. Obviously, the user will not want to create the corresponding input files by hand. For this purpose there are mesh-generating codes that create this data and require the user only to enter data related to boundary locations and number of elements desired. One such program, mesh.m, is presented in Appendix D. It creates all three of the data files just described. In addition, it provides the user with an easy method for assigning NPcode values to all boundary nodes.

6.3.6 Bandwidth and Bandwidth Reduction. A problem encountered with two-dimensional analyses is the numbering of the nodes to minimize the storage needed for the stiffness matrix. For our previous one-dimensional problems, the bandwidth was either 2 (for symmetric matrices) or 3 (for nonsymmetric matrices). With two-dimensional problems, the needed bandwidth will depend on the way the nodes have been numbered. Consider the two schemes shown in Fig. 6.5. For the first, the largest numerical difference between any two nodes associated with a single element is 3. This would require a bandwidth of 4 if the matrix were symmetric. For the second numbering scheme, the largest numerical difference is 8, which would require a bandwidth of 9. You might think that (1) the difference is not large and (2) it is easy to see how to number the nodes to get the smaller bandwidth. However, for complex meshes, (1) the difference between a good numbering scheme and a poor numbering scheme can result in a very large difference in bandwidth requirements, and (2) determining how to number the nodes to minimize the bandwidth can be difficult.

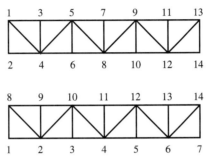

Figure 6.5. Two numbering schemes.

The optimal numbering scheme depends on the connectivity of the nodes, and this is completely specified by the NP array. Hence, computer codes are written that use this array as input to determine a new numbering scheme that will provide a narrow bandwidth. One such code, newnum.m, is described and given in Appendix D. The output from this code is the data file NWLD, which stands for new(old), that is, the new node number in terms of the old number. This array of numbers can be used as input to any finite element code to create a new numbering system that will produce a stiffness matrix having a narrow bandwidth. Whether program newnum.m or another code is used, NWLD must be in the working directory. The last entry in this code must be the bandwidth that the new numbering creates (see the example problem).

6.4 PROGRAM poisson.m

Program poisson.m is the finite element code for solving Eq. 6.1, Poisson's equation. In addition to poisson.m, you will need the following files in your working directory:

Input Data	User's INCLUDE	Supplied Functions
MESHo	INITIAL.m	sGAUSS.m
NODES	COEF.m	
NP		
NWLD		

MESHo is the data file that defines NUMNP, NUMEL, and NNPE. NODES is the data file that defines, for each node, its XORD, YORD, and NPcode values. NP is the connectivity array. NWLD gives the new nodal numbering for bandwidth reduction. INITIAL.m is the user's INCLUDE code for the initialization of variables to define the specific problem being analyzed. COEF.m is the user's INCLUDE code to define the parameters (coefficients) k_x, k_y, and Q. sGAUSS.m is the supplied function for solving the banded, symmetric equations by Gauss elimination.

Flow Chart

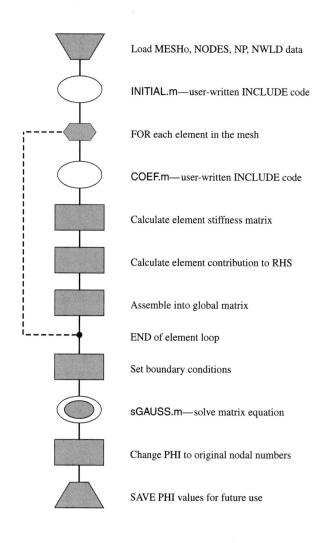

Load MESHo, NODES, NP, NWLD data

INITIAL.m—user-written INCLUDE code

FOR each element in the mesh

COEF.m—user-written INCLUDE code

Calculate element stiffness matrix

Calculate element contribution to RHS

Assemble into global matrix

END of element loop

Set boundary conditions

sGAUSS.m—solve matrix equation

Change PHI to original nodal numbers

SAVE PHI values for future use

6.4.2 Code

```
%------------------------
%   Program poisson.m
%------------------------



      clear
%------------------------
%   Load Data
%------------------------
      load MESHo    -ASCII
      load NODES    -ASCII
      load NP       -ASCII
      load NWLD     -ASCII




      NUMNP = MESHo(1);
      NUMEL = MESHo(2);
      NNPE  = MESHo(3);
      if NNPE ~= 3
        error('NNPE in MESHo  must equal 3')
      end


      for I=1:NUMNP;
          XORD(I)  =NODES(I,1);
          YORD(I)  =NODES(I,2);
          NPcode(I)=NODES(I,3);
      end
      IB = NWLD(NUMNP+1);

%   ----------------------
%   General Initialization
%   ----------------------
      for I=1:NUMNP;
         Q(I)=0;
         PHI(I)=0;
         NPBC(I)=0;
         for J=1:IB;
            SK(I,J)=0.0;
         end
      end

%   ----------------------
%   User's Initialization
%   ----------------------
      INITIAL
```

Poisson's equation

$$\frac{\partial}{\partial x}\left(R_x \frac{\partial \Phi}{\partial x}\right) + \frac{\partial}{\partial y}\left(R_y \frac{\partial \Phi}{\partial y}\right) = -Q$$

Load mesh and new numbering data. All four files must be in working directory.

File	Contents
MESHo	NUMNP, NUMEL, NNPE
NODES	XORD, YORD, NPBC
NP	NP array
NWLD	New nodal numbers
	Last entry is IB

Variable	Definition
NUMNP	Number of nodal points
NUMEL	Number of elements
NNPE	Number nodes per element

Error stop.
Program poisson.m is valid only for three-node triangular elements.

XORD	x coordinate of nodes
YORD	y coordinate of nodes
NPBC	Nodal point code to be used in initial.m
IB	Bandwidth for symmetric matrix

General initialization of all arrays

Include user's initialization code, INITIAL.m, for initialization peculiar to current problem being analyzed.

```
% -------------------------------
% Place Q in RHS, compact storage
% -------------------------------
  for I=1:NUMNP;
     RHS(NWLD(I))=Q(I);
  end

% -------------------------------------
% Formation of Finite Element Matrices
% Element by Element
% -------------------------------------
  for I=1:NUMEL;
    LMNT=I;
    XJ=XORD(NP(I,2))-XORD(NP(I,1));
    YJ=YORD(NP(I,2))-YORD(NP(I,1));
    XK=XORD(NP(I,3))-XORD(NP(I,1));
    YK=YORD(NP(I,3))-YORD(NP(I,1));

    XC=XORD(NP(I,1))+(XJ+XK)/3.0;
    YC=YORD(NP(I,1))+(YJ+YK)/3.0;

    VOL=(XJ*YK-XK*YJ)/2.0 ;
    if VOL < 0.0
     error('Element VOL is less ...
            than zero')
    end


% ----------------------------
% INCLUDE user's COEF.m code
% ----------------------------
  COEF

  COMM=1.0/(2.0*VOL);

  DNDX(1)=-(YK-YJ)*COMM;
  DNDX(2)=+(YK    )*COMM;
  DNDX(3)=-(YJ    )*COMM;
  DNDY(1)=+(XK-XJ)*COMM;
  DNDY(2)=-(XK    )*COMM;
  DNDY(3)=+(XJ    )*COMM;


  for J=1:3;
   Qe(J) = VOL*QVI/3.0;
    for K=1:3;
     S(J,K)=(DNDX(J)*RXI*DNDX(K) ...
            +DNDY(J)*RYI*DNDY(K))*VOL;
    end
  end
```

Create right-hand side with new nodal numbers.
Begin element-by-element integration.

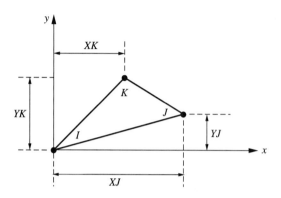

XC, YC   Global coordinates of element's centroid

VOL   Element's area: corresponds to volume for a unit thickness

ERROR if area is zero or negative

Include coefficients specified in user's COEF.m code: RXI = R_x, RYI = R_y, and QVI = Q.

$$DNDX = \left\{ \frac{\partial N}{\partial x} \right\}$$

$$DNDY = \left\{ \frac{\partial N}{\partial y} \right\}$$

$$\{Q\}_e = \int_{V_e} \{N\} Q \, dV$$

$$S(I, J) = \left\{ \frac{\partial N(I)}{\partial x} R_x \frac{\partial N(J)}{\partial x} + \frac{\partial N(I)}{\partial y} R_y \frac{\partial N(J)}{\partial y} \right\} V_e$$

$$[S] = \int_{V_e} [N']^T [R] [N'] \, dV$$

```
% --------------------------------
% Place in Global SK and Q matrices.
% Compact Storage
% --------------------------------

    for J=1:NNPE;
      newJ=NWLD(NP(I,J));
      RHS(newJ)=RHS(newJ)+Qe(J);
      for K=1:NNPE;
        newK=NWLD(NP(I,K));
        if newK >= newJ
          Kbnd=newK-newJ+1;
          SK(newJ,Kbnd)=...
            SK(newJ,Kbnd)+S(J,K);
        end
      end
    end


  end


% ---------------------------
% Specify boundary conditions
% ---------------------------
  for I=1:NUMNP;
    if NPBC(I) == 1
      Inew=NWLD(I);
      SK(Inew,1)=SK(Inew,1)*1.0E+10;
      RHS(Inew)=PHI(I)*SK(Inew,1);
    end
  end

% ------------------------------------
% Solution of Finite Element Equations
% ------------------------------------
  PHI = sGAUSS(SK,RHS,NUMNP,IB);

% ------------------------------------
% Place output in original numbering
% ------------------------------------
  for I=1:NUMNP;
     RHS(I) = PHI(NWLD(I));
  end
  for I=1:NUMNP;
     PHI(I) = RHS(I);
  end

  save PHI PHI -ASCII
```

Place element stiffness matrix S(I,J) into global stiffness matrix using compact storage for symmetric, banded matrices.

NP(I,J)	= Node *J* of element *I*
newJ	= New node number
NP(I,K)	= Node *K* of element *I*
newK	= New node number

Continue if (I,J) is on or above diagonal.

Kbnd	= Column of banded storage
SK	= Global stiffness matrix

END loop over elements.

If NPBC(I) = 1, then PHI is specified at this node.

Make diagonal large and adjust RHS to correspond to known Q.

Solve for Φ using sGAUSS.m, the equation solver for banded, symmetric matrices.

Place Φ values in order of original nodal numbering.

Save PHI for plotting and other purposes.

6.4.3 Test Problem. For our test problem, we again use the fact that whenever our finite element approximation can duplicate an exact solution, it will do so. The problem we have selected is defined as follows

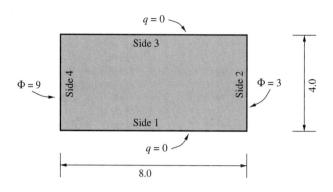

where Φ satisfies

$$\frac{\partial}{\partial x}\left[5\frac{\partial\Phi}{\partial x}\right] + \frac{\partial}{\partial y}\left[5\frac{\partial\Phi}{\partial y}\right] = 0$$

The exact solution to this problem is

$$\Phi = 9 - (6/8)x$$

where x is measured from the left side of the rectangle. Clearly, our piecewise linear approximation can duplicate this function; thus, poisson.m should produce the exact solution.

The first step in obtaining our finite element solution is to select a mesh. We choose a mesh with 16 elements and 15 evenly spaced nodes. The mesh and the corresponding data files are as follows

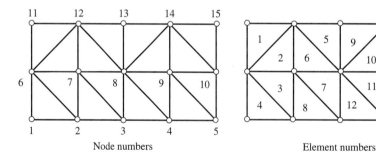

Node numbers   Element numbers

MESHo

15	% NUMNP
16	% NUMEL
3	% NNPE

NODES

% XORD	YORD	NPcode
0	0	4
2	0	0
4	0	0
6	0	0
8	0	2
0	2	4
2	2	0
4	2	0
6	2	0
8	2	2
0	4	4
2	4	0
4	4	0
6	4	0
8	4	2

Here we have defined our NPcode values to indicate the side of the mesh on which the node appears, and only on the sides where there is a need to specify a boundary condition that differs from the default values. We have chosen to number the sides counterclockwise, starting with the lower boundary. Thus, nodes 11, 6, and 1 are given NPcode numbers equal to 4, and nodes 15, 10, and 5 are given NPcode numbers equal to 2. However, there is no need to indicate that nodes 2, 3, and 4 are on side 1 or that nodes 12, 13, and 14 are on side 3, because there is no need to change any of the default values associated with these nodes. The important concept to understand is that the NPcode values are for your use and only your use in your INITIAL.m code. Use them there to identify the nodes for which you wish to assign boundary conditions that differ from the default boundary conditions. See the following example for the INITIAL.m code. Note, however, if you wish to assign boundary values in INITIAL.m without reference to the NPcode values, that is your option; you can simply give all nodes an NPcode value of zero and use, for example, the nodal coordinates to determine what boundary conditions to assign.

NP

```
%----------------
          6    12    11
          6     7    12
          6     2     7
          1     2     6
          8    13    12
          7     8    12
          7     3     8
          2     3     7
          8    14    13
          8     9    14
          8     4     9
          8     3     4
         14    10    15
         14     9    10
          9     5    10
          4     5     9
%--------------------
```

NWLD

```
%--------------
           3
           6
           9
          12
          15
           2
           5
           8
          11
          14
           1
           4
           7
          10
          13
           5    %   IB
%-----------------
```

The NWLD values have been assigned to make the numbering begin at the upper left corner and increase downward, similar to the example given earlier. Sketch the mesh with the new numbers and verify the NWLD data as well as the IB value shown. Keep in mind that only **poisson.m** uses the new numbers, not the user. In all data files, and in INITIAL.m, the original node numbers should be used.

The user INCLUDE codes for the test problem are:

INITIAL.m

```
%----------------------------
% Specify boundary conditions
%----------------------------
  for I=1:NUMNP
    if NPcode(I) == 2
      NPBC(I) = 1;
      PHI(I)  = 3;
      Q(I)    = 0;
    elseif NPcode(I) == 4
      NPBC(I) = 1;
      PHI(I)  = 9;
      Q(I)    = 0;
    end
  end
```

COEF.m

```
%===============================
%
%   Coefficients for poisson.m
%
%   RXI = k_x
%   RYI = k_Y
%   QVI = Q
%
%===============================

            RXI = 5.0;
            RYI = 5.0;
            QVI = 0.0;
```

In INITIAL.m, we search through all NPcode values to identify those equal to 2 (node is on side 2) and equal to 4 (node is on side 4). We again emphasize that these numbers simply point out to the user what the user wants them to point out. We could have used 3 and 9 to indicate the specified PHI values for these nodes and have accomplished the very same purpose.

The test problem has constant coefficients; hence, our COEF.m code is fairly simple. In problems with more complicated coefficients, for example, those that depend on the x and y coordinates, functions can be written in this code. Keep in mind when you write this code that you have use of all variables that have been calculated in poisson.m up to the line where COEF appears. This includes the coordinates of the element's centroid, XC and YC, as well as the gradients of the shape functions and other quantities.

At the conclusion of the analysis, the resulting Φ values are saved in your working directory in a file named PHI. For the test problem, it will contain the exact answers, which are

$$9.0 \quad 7.5 \quad 6.0 \quad 4.5 \quad 3.0$$
$$9.0 \quad 7.5 \quad 6.0 \quad 4.5 \quad 3.0$$
$$9.0 \quad 7.5 \quad 6.0 \quad 4.5 \quad 3.0$$

shown in the order corresponding to the mesh nodes.

Three auxiliary codes can be used in conjunction with poisson.m to help prepare input data and to help analyze the solution. They can be found in Appendix D and are as follows

mesh.m	Generates mesh files: MESHo, NODES, NP
newnum.m	Generates data file NWLD
topo.m	Creates plots of PHI values

EXERCISES

Study Problems

S1. Trace through the derivation leading to the weak form of Poisson's equation and determine if all the steps are valid if k_x and k_y are functions of x and y.

S2. Start with Eq. 6.11 and use the rules of variational calculus to arrive at the governing equation and boundary condition given by Eqs. 6.1 and 6.2.

S3. Complete the global SK matrix presented in the text.

S4. Show that Eq. 6.38 is correct.

S5. Explain why the bandwidth of the large SK matrix will be one greater than the largest difference between any two nodal point numbers in a single element.

S6. Consider how you could approximate $Q(x, y)$ by a function using nodal point values. How could you then make use of this representation to perform the integration for $\{Q\}$?

S7. In program poisson.m,
   (a) Explain in words the meaning of XORD(NP(I,3)).
   (b) Verify the equations for XC and YC.
   (c) Verify the equation for VOL.
   (d) Verify the equation for DNDY(1).
   (e) Determine the number of arithmetic operations (additions, subtractions, multiplications, and divisions) that take place within the two nested loops used to create the element stiffness matrix S(J,K).

S8. For the test problem, draw the mesh and number the nodes according to the NWLD data file.

S9. For the new numbering of the nodes as given in NWLD for the test problem, verify that the value for IB given is correct.

Numerical Experiments and Code Development

N1. At the end of poisson.m use the newly calculated values of PHI to calculate the {Q} array. Note

$$[K]\{\Phi\} = \{Q\}$$

so divide the diagonal terms that have been multiplied by the large factor to regain their original value for the above multiplication. Also remember that SK is in symmetric, banded storage.

N2. Change the boundary conditions for the test problem to specify PHI on the bottom surface of the rectangle to be 7.0 and on the top surface to be −5.0. For the two vertical sides, specify no flux. Run poisson.m and verify that your answer is correct.

N3. Change the boundary conditions for the test problem and specify PHI on all sides according to

$$\Phi = 3x - 7y$$

Run poisson.m and verify that the exact answer is obtained.

N4. Run the preceding problem, but change all nodal point coordinates so as to make the mesh represent a rather nondescript shape.

N5. Change the boundary conditions for the test problem to specify PHI values on side 4 to be 25 and the flux on side 2 to be −33. Specify zero flux on the two other surfaces. Verify that poisson.m gives the exact answer.

N6. Change the NWLD array to some other numbering and calculate the IB value. Rerun the test problem to verify that the new numbering gives the same results.

CHAPTER 7

POISSON'S EQUATION: APPLICATIONS

We now consider applications of Poisson's equation to (1) flow through porous media, (2) heat transfer by conduction, (3) flow of ideal (nonviscous) fluids, (4) torsion of noncircular shafts, (5) parallel flow of viscous fluids, and (6) problems in electrostatics. We will pay particular attention to the dimensions (force, length, time, and temperature) of terms in the governing equations and in the corresponding finite element approximations.

In these applications, Poisson's equation will represent the merging of two first-order differential equations. For several of the applications, the two first-order equations will be (1) a conservation law and (2) a constitutive equation. In these problems, the region of analysis is considered a control volume; hence, our volume designation in the previous chapter will be appropriate. However, for the torsion problem and the problem of parallel flow of viscous fluids, it will be more appropriate to think of the region of analysis as an area.

The first example will serve as a prototype application. We will discuss it in detail to pave the way for the examples that follow.

7.1 FLOW THROUGH POROUS MEDIA

Figure 7.1a illustrates a typical problem associated with flow in porous media. The concern is seepage under the dam and uplift pressure on the dam. The pressure upstream and downstream is shown to be hydrostatic; hence, the boundaries are assumed to be sufficiently far removed from the dam that no flow crosses them. The dam itself is considered long in the direction perpendicular to the page, making it possible to assume a state of two-dimensional flow under the dam. The lower boundary is assumed to be impervious rock that prevents any flow across that boundary.

Figure 7.1b shows the piezometric head, which is the dependent function to be obtained through the solution of Poisson's equation. The sheet pile under the dam produces a sudden decrease in this head, thus reducing the uplift pressure on the dam. Figure 7.1c illustrates a possible finite element mesh for the approximation of the Φ surface.

To formulate the governing equation for the flow under the dam, it is first necessary to understand Darcy's law. Figure 7.2 depicts Darcy's original experiment. A pipe filled with a porous material has water flowing through it due to a change in the piezometric head, H, from one end of the pipe to the other end. The piezometric head is the elevation to which water will rise in a tube placed at the point in question. As shown in Fig. 7.2, the head is made up of two distinct parts: the elevation head and the pressure head. Darcy found that the average velocity in the pipe is proportional to the change in the piezometric head; hence, in the limit,

$$u = -R\frac{dH}{dS} \tag{7.1}$$

where R is the coefficient of permeability. This equation can be extended to two-dimensional flow to give us

$$u_x = -R_x\frac{\partial H}{\partial x}$$
$$u_y = -R_y\frac{\partial H}{\partial y} \tag{7.2}$$

Here we have assumed that the directions x and y are the principal directions of hydraulic conductivity. When this is not the case, there can be a flow created perpendicular to the gradient. This can be accounted for by treating the hydraulic conductivity as a second-rank tensor. In what follows, it is assumed that the axes do line up with the principal directions of the hydraulic conductivity. In many applications, the material is isotropic; hence, $R_x = R_y$ and the above equations hold regardless of the orientation of the axes.

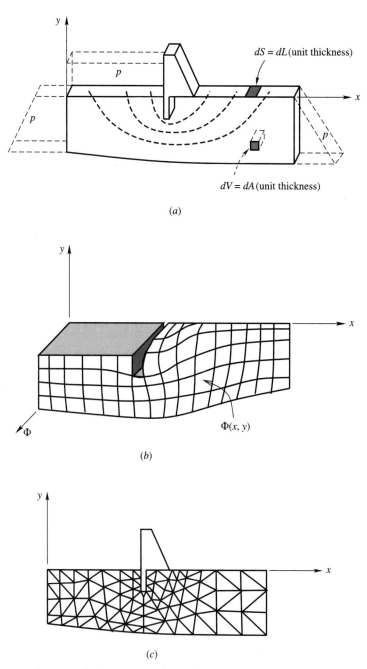

Figure 7.1. Seepage under a dam.

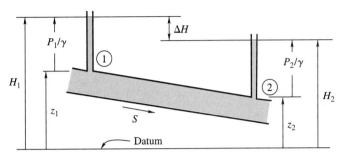

Figure 7.2. Darcy's experiment.

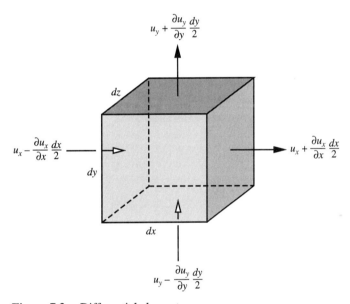

Figure 7.3. Differential element.

The conservation principle to be enforced applies to the volume (mass) of the incompressible fluid. Hence, the total flow leaving through the surfaces of the differential element shown in Fig. 7.3 must equal that entering plus the internal sources. We have, therefore,

$$\left[\frac{\partial u_x}{\partial x}dx \right] dy \, dz + \left[\frac{\partial u_y}{\partial y}dy \right] dx \, dz = Q \, dx \, dy \, dz \tag{7.3}$$

where u_x and u_y are the components of velocity at the centroid of the differential element. Upon dividing by $dx\,dy\,dz$, we obtain

$$\frac{\partial u_x}{\partial x} + \frac{\partial u_y}{\partial y} = Q$$

$$\boxed{\frac{L}{T}\frac{1}{L}} \qquad \boxed{\frac{L}{T}\frac{1}{L}} \qquad \boxed{\frac{L^3}{T}\frac{1}{L^3}}$$

(7.4)

where L and T are the dimensions of length and time, respectively.

Substitution of Darcy's law now gives us

$$\frac{\partial}{\partial x}\left[-R_x\frac{\partial H}{\partial x}\right] + \frac{\partial}{\partial y}\left[-R_y\frac{\partial H}{\partial y}\right] = +Q$$

$$\frac{\partial}{\partial x}\left[R_x\frac{\partial H}{\partial x}\right] + \frac{\partial}{\partial y}\left[R_y\frac{\partial H}{\partial y}\right] = -Q$$

(7.5)

$$\boxed{\frac{1}{L}\frac{L}{T}\frac{L}{L}} \qquad \boxed{\frac{1}{L}\frac{L}{T}\frac{L}{L}} \qquad \boxed{\frac{L^3}{T}\frac{1}{L^3}}$$

$$\boxed{\frac{1}{T}} \qquad \boxed{\frac{1}{T}} \qquad \boxed{\frac{1}{T}}$$

where R_x and R_y are the coefficients of hydraulic conductivity and Q is a source per unit volume. We consider next a segment of surface as shown in Fig. 7.4. The component of velocity normal and outward to the surface is found by taking the scalar product of the outward unit normal vector to the surface with the velocity vector. This gives us

$$u_n = u_x n_x + u_y n_y$$

(7.6)

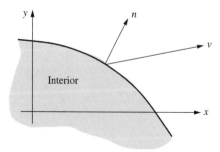

Figure 7.4.   Segment of surface.

Hence,

$$u_n = \left(-R_x \frac{\partial H}{\partial x}\right) n_x + \left(-R_y \frac{\partial H}{\partial y}\right) n_y = \boxed{\text{Flow leaving}} \tag{7.7}$$

After a change in sign, we obtain

$$R_x \frac{\partial H}{\partial x} n_x \quad + \quad R_y \frac{\partial H}{\partial y} n_y \quad = \quad \boxed{\text{Flow entering}}$$

$$\boxed{\dfrac{L}{T} \quad \dfrac{L}{L} \quad 1} \quad \boxed{\dfrac{L}{T} \quad \dfrac{L}{L} \quad 1} \quad \boxed{\dfrac{L^3}{T} \dfrac{1}{L^2}} \tag{7.8}$$

$$\boxed{\dfrac{\text{(Vol. of fluid)}}{\text{(Unit of time)}} \dfrac{(1)}{\text{(Unit surface area)}}}$$

$$\boxed{\begin{array}{c}\text{Flow entering surface}\\ \text{per unit surface area}\end{array}}$$

Now consider our finite element approximation applied to this problem. The element stiffness matrix would have the following dimensions:

$$[\,K\,]_e = \int_{V_e} [N']^T \quad [R] \quad [N'] \quad dV$$

$$\boxed{\dfrac{L^2}{T}} \qquad \boxed{\dfrac{1}{L} \ \Big| \ \dfrac{L}{T} \ \Big| \ \dfrac{1}{L} \ \Big| \ L^3} \tag{7.9}$$

The right-hand side, $\{Q\}$, is made up of terms coming from the distributed source $Q(x, y)$ and the surface flux $q(S)$. Its dimensions are as follows:

$$\{Q\}_e = \int_{V_e} \{N\} \quad Q(x, y) \quad dV$$

$$\boxed{\dfrac{L^3}{T}} \qquad \boxed{1 \ \Big| \ \dfrac{1}{T} \ \Big| \ L^3} \tag{7.10}$$

$$\{Q\}_s = \int_{S_s} \{N_s\} \quad q(s) \quad ds$$

(7.11)

$$\boxed{\dfrac{L^3}{T}} \qquad \boxed{1 \;\Big|\; \dfrac{L}{T} \;\Big|\; L^2}$$

Our finite element equation thus has the following dimensions:

$$[K] \quad \{H\} \quad = \quad \{Q\}$$

(7.12)

$$\boxed{\dfrac{L^2}{T} \;\Big|\; L} \qquad \boxed{\dfrac{L^3}{T}}$$

where the H's are the nodal point values for the piezometric (total) head, and the Q's are the nodal point sources from both volume and surface integrations.

We can place a physical significance on the terms of the K matrix. Consider the two-element mesh shown in Fig. 7.5 and the corresponding matrix equation

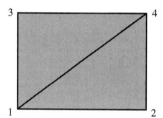

Figure 7.5. A two element mesh.

$$
\begin{array}{ccc}
[K] & \{\mathrm{HEAD}\} & \{\mathrm{SOURCE}\}
\end{array}
$$

$$
\begin{bmatrix}
(1,1) & (1,2) & (1,3) & (1,4) \\
(2,1) & (2,2) & (2,3) & (2,4) \\
(3,1) & (3,2) & (3,3) & (3,4) \\
(4,1) & (4,2) & (4,3) & (4,4)
\end{bmatrix}
\begin{Bmatrix}
0 \\ 1 \\ 0 \\ 0
\end{Bmatrix}
=
\begin{Bmatrix}
(1,2) \\ (2,2) \\ (3,2) \\ (4,2)
\end{Bmatrix}
=
\begin{Bmatrix}
Q(1) \\ Q(2) \\ Q(3) \\ Q(4)
\end{Bmatrix}
$$

We see that a unit head at a node J, with all other nodes having a zero head, produces a source at any other node I equal to $K(I, J)$.

We also observe that because K is a symmetric matrix, $K(I, J) = K(J, I)$, we have a reciprocal theorem for porous flow. That is, the effect of a change in head at point A on the flow at point B is the same as the effect of an equal change in head at point B on flow at point A.

Finally, we note that because the flow is incompressible, the total sources must sum to zero. Hence, we see from the preceding equation that the sum of any column must equal zero. Because of symmetry, this is also true of the rows.

7.2 STEADY-STATE HEAT TRANSFER BY CONDUCTION

A steady-state temperature field is governed by the law of conservation of energy which requires that the total heat entering and leaving a unit volume sum to zero. The heat entering through the sides of a unit volume is governed by Fourier's law, which for one dimension is

$$q \quad = \quad -R \quad \frac{d\Phi}{dS}$$

$\dfrac{FL}{TL^2}$	$\dfrac{FL}{TL\Theta}$	$\dfrac{\Theta}{L}$
Heat flow per unit area	Thermal conductivity	Temperature gradient

$$(7.13)$$

Here F, L, T, and Θ are the dimensions of force, length, time, and temperature, respectively.

The heat entering a unit volume is the sum of that entering through its sides and the interior heat source. Thus, for a two-dimensional unit volume, we can sum

$$\frac{\partial}{\partial x}\left[R_x\frac{\partial\Phi}{\partial x}\right] + \frac{\partial}{\partial y}\left[R_y\frac{\partial\Phi}{\partial y}\right] = \boxed{\text{Heat entering through surfaces of unit volume}}$$

$$Q(x, y) = \boxed{\text{Interior heat source per unit volume}}$$

to obtain

$$\frac{\partial}{\partial x}\left[R_x \quad \frac{\partial\Phi}{\partial x}\right] + \frac{\partial}{\partial y}\left[R_y \quad \frac{\partial\Phi}{\partial y}\right] + Q(x, y) = 0$$

$\dfrac{1}{L}$	$\dfrac{FL}{TL\Theta}$	$\dfrac{\Theta}{L}$	$\dfrac{1}{L}$	$\dfrac{FL}{TL\Theta}$	$\dfrac{\Theta}{L}$	$\dfrac{FL}{L^3T}$

$$(7.14)$$

The finite element equations and dimensions are

$$\Phi(x, y) = \lfloor N \rfloor \{\Phi\}_e \quad \text{and} \quad \{\nabla\Phi\} = [N'] \{\Phi\}_e$$

| Θ | 1 | Θ | | $\dfrac{\Theta}{L}$ | $\dfrac{1}{L}$ | Θ |

(7.15)

$$[K]_e = \int_{V_e} [N']^T [R] [N'] \, dV$$

| $\dfrac{FL}{\Theta T}$ | | $\dfrac{1}{L}$ | $\dfrac{FL}{TL\Theta}$ | $\dfrac{1}{L}$ | L^3 |

(7.16)

$$\{Q\}_e = \int_{V_e} \{N\} \, Q(x, y) \, dV$$

| $\dfrac{FL}{T}$ | | 1 | $\dfrac{FL}{TL^3}$ | L^3 |

(7.17)

$$\{Q\}_s = \int_{S_s} \{N_s\} \, q(S) \, dS$$

| $\dfrac{FL}{T}$ | | 1 | $\dfrac{FL}{TL^2}$ | L^2 |

(7.18)

Thus, our final finite element matrix equation has the following dimensions when applied to steady-state heat conduction:

$$[K] \quad \{\Phi\} = \{Q\}$$

| $\dfrac{FL}{\Theta T}$ | Θ | | $\dfrac{FL}{T}$ |

(7.19)

7.3 FLOW OF IDEAL FLUIDS

Potential Function. Let u and v be the x and y components of the velocity of a inviscid fluid. If the flow is irrotational, we have

$$\frac{\partial}{\partial y}(u) - \frac{\partial}{\partial x}(v) = 0 \tag{7.20}$$

$$\boxed{\frac{1}{L}\,\bigg|\,\frac{L}{T}} \qquad \boxed{\frac{1}{L}\,\bigg|\,\frac{L}{T}}$$

By defining a potential function Φ such that

$$u = \frac{\partial}{\partial x}(\Phi) \quad \text{and} \quad v = \frac{\partial}{\partial y}(\Phi) \tag{7.21}$$

$$\boxed{\frac{L}{T}} \quad \boxed{\frac{1}{L}\,\bigg|\,\frac{L^2}{T}} \qquad \boxed{\frac{L}{T}} \quad \boxed{\frac{1}{L}\,\bigg|\,\frac{L^2}{T}}$$

we guarantee that Eq. 7.20 is identically satisfied. If the flow is incompressible,

$$\frac{\partial u}{\partial x} + \frac{\partial v}{\partial y} = 0 \tag{7.22}$$

and substitution of the potential function gives us

$$\frac{\partial^2}{\partial x^2}[\Phi] + \frac{\partial^2}{\partial y^2}[\Phi] = 0 \tag{7.23}$$

$$\boxed{\frac{1}{L^2}\,\bigg|\,\frac{L^2}{T}} \qquad \boxed{\frac{1}{L^2}\,\bigg|\,\frac{L^2}{T}}$$

The boundary conditions for Φ are

$$\vec{u} \cdot \vec{n} = un_x + vn_y$$

$$= \frac{\partial}{\partial x}[\Phi]\,n_x + \frac{\partial}{\partial y}[\Phi]\,n_y = q \tag{7.24}$$

$$\boxed{\frac{1}{L}\,\bigg|\,\frac{L^2}{T}\,\bigg|\,1} \qquad \boxed{\frac{1}{L}\,\bigg|\,\frac{L^2}{T}\,\bigg|\,1} \qquad \boxed{\frac{L}{T}}$$

where q is the outward component of velocity.

Stream Function. In the preceding formulation, the two governing equations,

$$\frac{\partial u}{\partial y} - \frac{\partial v}{\partial x} = 0 \quad \text{Irrotational flow}$$

$$\frac{\partial u}{\partial x} + \frac{\partial v}{\partial y} = 0 \quad \text{Incompressible flow}$$

were satisfied by (1) defining a potential function that guaranteed that the flow was irrotational (i.e., that Eq. 7.20 was identically satisfied) and (2) requiring that the potential function satisfy Eq. 7.23, that is, requiring the flow field defined by the potential function to be incompressible.

The procedure can be reversed to obtain a second formulation. If we define Ψ such that the condition for incompressible flow is identically satisfied,

$$u = +\frac{\partial}{\partial y} \lfloor \Psi \rfloor \quad \text{and} \quad v = -\frac{\partial}{\partial x} \lfloor \Psi \rfloor$$

$$\boxed{\frac{L}{T}} \quad \boxed{\frac{1}{L} \,\bigg|\, \frac{L^2}{T}} \qquad\qquad \boxed{\frac{L}{T}} \quad \boxed{\frac{1}{L} \,\bigg|\, \frac{L^2}{T}}$$

$$(7.25)$$

then the equation specifying irrotational flow is

$$\frac{\partial^2}{\partial x^2} \lfloor \Psi \rfloor + \frac{\partial^2}{\partial y^2} \lfloor \Psi \rfloor = 0$$

$$\boxed{\frac{1}{L^2} \,\bigg|\, \frac{L^2}{T}} \quad \boxed{\frac{1}{L^2} \,\bigg|\, \frac{L^2}{T}}$$

$$(7.26)$$

This is the same governing equation as was found for the potential function. The difference between the two methods lies in the specification of the boundary conditions. For the stream function, specification of the derivative tangent to the boundary is equivalent to specifying the velocity across the boundary. For the potential function Φ, specification normal to the boundary is equivalent to specifying the velocity across the boundary.

Because the potential function and the stream function have the same dimensions, there is no distinction between their finite element approximations except for the boundary conditions. If we now use Φ to represent either the potential function or the the stream function, we can write for both

$$\Phi(x, y) = \lfloor N \rfloor \{\Phi\}$$

$$\boxed{\frac{L^2}{T}} \qquad \boxed{1} \quad \boxed{\frac{L^2}{T}}$$

$$(7.27)$$

$$[k]_e = \int_{V_e} [N']^T \; [N'] \; dv$$

(7.28)

$$\boxed{L} \qquad \boxed{\dfrac{1}{L}} \; \boxed{\dfrac{1}{L}} \; \boxed{L^3}$$

For the stream function approach, Φ is known along all boundaries. For the potential function approach, the normal derivative of Φ is specified; hence,

$$\{Q\}_s = \int_{S_s} \{N\} \; q \; dS$$

(7.29)

$$\boxed{\dfrac{L^3}{T}} \qquad \boxed{1} \; \boxed{\dfrac{L^3}{TL^2}} \; \boxed{L^2}$$

where

$$q = \dfrac{d}{dn} [\Phi]$$

(7.30)

$$\boxed{\dfrac{L}{T}} \qquad \boxed{\dfrac{1}{L}} \; \boxed{\dfrac{L^2}{T}}$$

The global matrix equation for both approximations is

$$[K] \; \{\Phi\} = \{Q\}$$

(7.31)

$$\boxed{L \; \Big| \; \dfrac{L^2}{T}} \qquad \boxed{\dfrac{L^3}{T}}$$

7.4 TORSION OF PRISMATIC BARS

Refer to Fig. 7.6 for the following definitions and equations:

$$\alpha = \text{angle of twist}$$

$$\Theta = \alpha/L = \text{angle of twist per unit length of rod}$$

$$\left.\begin{array}{c} \tau_{zx} \\ \tau_{zy} \end{array}\right\} = \text{shear stress components on } z \text{ face}$$

$$\Phi = \text{stress function defined such that}$$

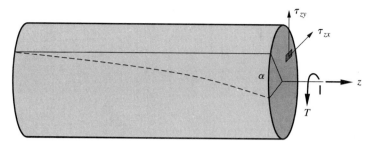

Figure 7.6. Torsion of a cylindrical shaft.

$$\tau_{zx} \;=\; \frac{\partial}{\partial y} \;\Phi \quad \text{and} \quad \tau_{zy} \;=\; -\frac{\partial}{\partial x} \;\Phi$$

(7.32)

| $\dfrac{F}{L^2}$ | | $\dfrac{1}{L}$ | $\dfrac{F}{L}$ | | $\dfrac{F}{L^2}$ | | $\dfrac{1}{L}$ | $\dfrac{F}{L}$ |

The preceding definition of a potential function for the stress will create stress fields that satisfy the equations of equilibrium. Of all stress fields generated from the potential function, those that produce compatible strains must satisfy

$$\frac{\partial^2}{\partial x^2}\;[\Phi] \;+\; \frac{\partial^2}{\partial y^2}\;[\Phi] \;=\; -2\;G\;\Theta$$

(7.33)

| $\dfrac{1}{L^2}$ | $\dfrac{F}{L}$ | | $\dfrac{1}{L^2}$ | $\dfrac{F}{L}$ | | $\dfrac{F}{L^2}$ | $\dfrac{1}{L}$ |

where G is the material's modulus of rigidity. The boundary conditions are

$$\Phi = \text{constant on all boundaries}$$

(7.34)

and

$$M_z \;=\; 2\int_A \Phi \; dA$$

(7.35)

| FL | | $\dfrac{F}{L}$ | L^2 |

The governing equation will be written in the following form:

$$\frac{\partial}{\partial x}\left(\frac{1}{G}\frac{\partial \Phi}{\partial x}\right) + \frac{\partial}{\partial y}\left(\frac{1}{G}\frac{\partial \Phi}{\partial y}\right) = -2\,\Theta \tag{7.36}$$

$$\boxed{\begin{array}{|c|c|c|} \hline \dfrac{1}{L} & \dfrac{L^2}{F} & \dfrac{F}{L^2} \\ \hline \end{array}} \quad \boxed{\begin{array}{|c|c|c|} \hline \dfrac{1}{L} & \dfrac{L^2}{F} & \dfrac{F}{L^2} \\ \hline \end{array}} \quad \boxed{\dfrac{1}{L}}$$

The finite element equations are

$$[K]_e = \int_{A_e} [N']^T \ [R] \ [N'] \ dA \tag{7.37}$$

$$\boxed{\dfrac{L^2}{F}} \quad \boxed{\begin{array}{|c|c|c|c|} \hline \dfrac{1}{L} & \dfrac{L^2}{F} & \dfrac{1}{L} & L^2 \\ \hline \end{array}}$$

$$\{Q\}_e = \int_{A_e} \{N\} \ Q \ dA \tag{7.38}$$

$$\boxed{L} \quad \boxed{\begin{array}{|c|c|c|} \hline 1 & \dfrac{1}{L} & L^2 \\ \hline \end{array}}$$

where $Q = -2\Theta$. Our final finite element equation applied to the torsion problem thus has the following dimensions:

$$[K] \ \{\Phi\} = \{Q\} \tag{7.39}$$

$$\boxed{\begin{array}{|c|c|} \hline \dfrac{L^2}{F} & \dfrac{F}{L} \\ \hline \end{array}} \quad \boxed{L}$$

Note that in the preceding equations we have used dA rather than dV because the physical interpretation of these equations does not imply a unit thickness associated with the plane area used for the analysis.

7.5 PARALLEL FLOW OF VISCOUS FLUIDS

The flow of viscous fluids in pipes of constant cross-section is parallel; hence, there is only one unknown component of velocity. In addition, for cross-sections far removed from the entrance, the velocity remains the same from one cross-section to the next. Thus, the problem reduces to a two-dimensional problem with one dependent variable. Such a flow is illustrated in Fig. 7.7 for a pipe of rectangular cross-section.

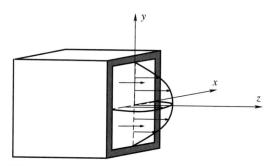

Figure 7.7. Parallel flow of a viscous fluid in a pipe.

The governing equation is the Navier-Stokes equation:

$$\rho v_k \frac{\partial v_i}{\partial x_k} = -\frac{\partial p}{\partial x_i} + \frac{\partial}{\partial x_k}\left\{\mu\left(\frac{\partial v_i}{\partial x_k} + \frac{\partial v_k}{\partial x_i}\right)\right\} \tag{7.40}$$

where

$$\rho = \text{density of fluid}$$

$$v_k = k\text{th component of velocity}$$

$$x_k = k\text{th coordinate}$$

$$p = \text{pressure}$$

$$\mu = \text{viscosity of the fluid}$$

For parallel flow,

$$v_x = v_y = 0$$

$$v_z = v_z(x, y)$$

Substitution of these velocities into the x and y components of the Navier-Stokes equation shows that they are identically satisfied if the pressure is a function only of z. Substitution into the z component gives

$$\frac{\partial}{\partial x}\left(\mu\frac{\partial u_z}{\partial x}\right) + \frac{\partial}{\partial y}\left(\mu\frac{\partial u_z}{\partial y}\right) = -\frac{\partial p}{\partial z} \tag{7.41}$$

where the dimensions of each term are

$$\frac{\partial}{\partial x}\left\{\mu \quad \frac{\partial u_z}{\partial x}\right\} + \frac{\partial}{\partial y}\left\{\mu \quad \frac{\partial u_z}{\partial y}\right\} = -\frac{\partial p}{\partial z}$$

$\frac{1}{L}$	$\frac{FT}{L^2}$	$\frac{L}{TL}$	$\frac{1}{L}$	$\frac{FT}{L^2}$	$\frac{L}{TL}$	$\frac{F}{L^2 L}$

The finite element equations are

$$[K]_e = \int_{A_e} [N']^T \quad [R] \quad [N'] \quad dA$$

(7.42)

$$\boxed{\frac{FT}{L^2}} \qquad \boxed{\frac{1}{L}} \ \boxed{\frac{FT}{L^2}} \ \boxed{\frac{1}{L}} \ \boxed{L^2}$$

$$\{Q\}_e = \int_{A_e} \{N\} \quad \frac{\partial p}{\partial z} \quad dA$$

(7.43)

$$\boxed{\frac{F}{L}} \qquad \boxed{1} \ \boxed{\frac{F}{L^2 L}} \ \boxed{L^2}$$

The final finite element equation is

$$[K] \quad \{\Phi\} = \{Q\}$$

(7.44)

$$\boxed{\frac{FT}{L^2}} \ \boxed{\frac{L}{T}} \qquad \boxed{\frac{F}{L}}$$

where Φ is the nodal velocity in the z direction. Note that dA has been used rather than dV because the physical interpretation of all equations is that they apply over an area rather than a volume of unit thickness. Because the cross-section of the pipe does not change along its length, the pressure gradient is constant throughout the pipe. When unknown, it can be taken as unity and the resulting velocity would be the velocity per unit pressure drop.

Before leaving, it is necessary to examine the boundary conditions associated with this problem. For viscous fluids, the velocity is zero along the sides of the pipe. If a boundary represents a plane of symmetry, then the gradient of the velocity is zero normal to the boundary. For this case, the surface integral resulting from the integration by parts is zero, and there is no need to add additional terms to the right-hand side.

7.6 ELECTROSTATICS

The application of Poisson's equation to electrostatics can be shown by considering two fundamental laws. In their differential form they are Gauss's law,

$$\mathbf{\nabla} \cdot \mathbf{D} = \rho_T$$

$$\frac{\partial}{\partial x} \quad (D_x) + \frac{\partial}{\partial y} \quad (D_y) = \rho_T$$

(7.45)

$$\boxed{\frac{1}{L}} \ \boxed{\frac{Q}{L^2}} \qquad \boxed{\frac{1}{L}} \ \boxed{\frac{Q}{L^2}} \qquad \boxed{\frac{Q}{L^3}}$$

and Faraday's law,

$$\nabla \times \mathbf{E} = 0$$

$$\frac{\partial}{\partial x}\left(E_y\right) - \frac{\partial}{\partial y}(E_x) = 0$$

| $\dfrac{1}{L}$ | $\dfrac{F}{Q}$ |

| $\dfrac{1}{L}$ | $\dfrac{F}{Q}$ |

(7.46)

In these equations,

$$\mathbf{D} = \text{electric flux density}$$

$$\rho_T = \text{free charge density}$$

$$\mathbf{E} = \text{electric field intensity}$$

and F, L, and Q represent the dimensions of force, length, and charge, respectively.

Equation 7.46 is identically satisfied by any electric intensity field derived from a potential function, V, such that

$$\mathbf{E} = -\nabla V$$

$$E_x = -\frac{\partial}{\partial x}(V) \qquad E_y = -\frac{\partial}{\partial y}(V)$$

| $\dfrac{F}{Q}$ |

| $\dfrac{1}{L}$ | $\dfrac{FL}{Q}$ |

| $\dfrac{F}{Q}$ |

| $\dfrac{1}{L}$ | $\dfrac{FL}{Q}$ |

(7.47)

where V is referred to as the *electric potential.*

The electric flux density and electric field intensity in dielectric materials are related by

$$\mathbf{D} = \epsilon \mathbf{E}$$

$$D_x = \epsilon \, E_x \qquad D_y = \epsilon \, E_y$$

| $\dfrac{Q}{L^2}$ |

| $\dfrac{Q^2}{FL^2}$ | $\dfrac{F}{Q}$ |

| $\dfrac{Q}{L^2}$ |

| $\dfrac{Q^2}{FL^2}$ | $\dfrac{F}{Q}$ |

(7.48)

where ϵ is the material's dielectric constant. In this case, we have

$$\nabla \cdot (\epsilon \nabla V) = -\rho_T$$

$$\frac{\partial}{\partial x}\left(\epsilon \frac{\partial V}{\partial x}\right) + \frac{\partial}{\partial y}\left(\epsilon \frac{\partial V}{\partial y}\right) = -\rho_T \tag{7.49}$$

$\dfrac{1}{L}$	$\dfrac{Q^2}{FL^2}$	$\dfrac{FL}{QL}$

$\dfrac{1}{L}$	$\dfrac{Q^2}{FL^2}$	$\dfrac{FL}{QL}$

$\dfrac{Q}{L^3}$

To complete the problem statement, either V or q must be specified at all points on the boundary, where

$$q = \epsilon \frac{\partial V}{\partial n} \tag{7.50}$$

$\dfrac{Q}{L^2}$

$\dfrac{Q^2}{L^2 F}$	$\dfrac{FL}{QL}$

When the preceding equations are placed in their weak form and a finite element approximation used for V, integration over an element volume gives us

$$[K]_e = \int_{V_e} [N']^T [R] [N'] \, dV \tag{7.51}$$

$\dfrac{Q^2}{FL}$

$\dfrac{1}{L}$	$\dfrac{Q^2}{FL^2}$	$\dfrac{1}{L}$	L^3

$$\{Q\}_e = \int_{V_e} \{N\} \, \epsilon \, dV \tag{7.52}$$

Q

1	$\dfrac{Q}{L^3}$	L^3

and integration over a surface segment gives

$$\{Q\}_s = \int_{S_s} \{N_s\} \, \epsilon \, \frac{\partial V}{\partial n} \, dS \tag{7.53}$$

Q

1	$\dfrac{Q^2}{FL^2}$	$\dfrac{FL}{LQ}$	L^2

Assembly into the global stiffness matrix and right-hand side gives us our final finite element equation:

$$[K] \quad \{V\} \quad = \quad \{Q\}$$

(7.54)

$\dfrac{Q^2}{FL}$	$\dfrac{FL}{Q}$

$\boxed{Q}$

A simplification can be made by noting that Q^2 has the dimensions of FL^2. With this, we have

$$[K] \quad \{V\} \quad = \quad \{Q\}$$

(7.55)

L	$\dfrac{Q}{L}$

$\boxed{Q}$

Note that this substitution also shows us that the material's dielectric constant, ϵ, is a dimensionless parameter.

EXERCISES

Study Problems

S1. In each application, determine the dimensions of the functional J.

S2. In the chapter applications, sudden changes in materials can be taken into account by changing the values of the parameters R_x and R_y. How does this affect the integration by parts used to arrive at the finite element equations?

S3. For each application, is there a meaningful reciprocal theorem? If so, describe in words what it is.

Numerical Experiments and Code Development

N1. For problems where Q is a function of x and y, represent it by its nodal point values and read the values in as input data. Then use the nodal values in your COEF.m code to determine QVI, the value of Q at the centroid of the element.

N2. Write a program that will load the mesh data files and the PHI file obtained from poisson.m and then determine the gradient vector at the centroid of each element. Plot these vectors as arrows, based on a scale proportional to an average mesh spacing (i.e., do not have arrows that overlap a great deal by spanning several elements).

Projects

General instructions: For each numerical application, check the accuracy of your solution by conducting an additional analysis with a finer mesh. If this produces a change in your solution greater than the accuracy you desire, then an additional analysis should be conducted with an even finer mesh. This sequence should be continued until the accuracy you desire is obtained. Whenever possible, symmetry in the physical problem should be taken into account to save computer time.

P1. *Thermal distribution in a flat slab.* Use the following region to study the temperature distribution under the following conditions:

	Boundary Temperatures				
Case	A	B	C	D	Q
1	?	+1.0	−1.0	?	0.0
2	+1.0	−1.0	+1.0	−1.0	0.0
3	0.0	+1.0	−1.0	0.0	0.0
4	0.0	0.0	0.0	0.0	1.0

where the question mark indicates an insulated boundary and Q is the distributed heat source. In all cases assume the upper and lower boundaries are insulated. Assume that the units of length, energy, and temperature of the values shown are consistent with a unit value for the coefficient of thermal conductivity.

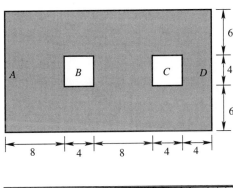

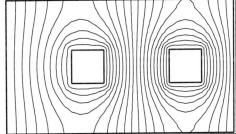

Isotherms for Case 3.

P2. *Flow of a perfect fluid around a cylinder.* A channel of thickness h has a cylindrical obstruction in its center. The radius of the cylinder is $R = h/4$. Determine the flow of an ideal fluid that enters at the left with a uniform unit velocity and exits at the right, also with a uniform velocity. Use both the potential flow formulation and the streamline formulation. Conduct several analyses to determine the effect of L on the solution. At what value of L does continued extension of the boundary appear to have no effect?

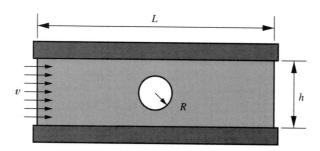

P3. *Torsion of rods with multiply connected cross-sections.* For shafts of simply connected, noncircular cross-sections, the boundary value for Φ may be selected arbitrarily. However, for multiply connected cross-sections, it is permissible to assign Φ arbitrarily on only one of the boundaries. For the other boundaries, Φ must be determined using equations that ensure a single-valued warping function.

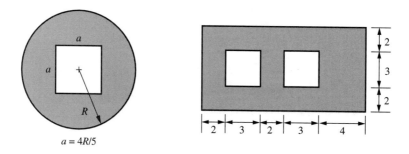

This, however, can be avoided by assuming the cross-section to be simply connected but made of more than one material. If interior holes are filled with a material that has a modulus of rigidity several orders of magnitude less than the actual material, there should be very little effect on the total rigidity of the shaft. One advantage of the finite element method is that it can easily accommodate such abrupt changes in material properties.

Use this approach to determine the torsional rigidity of the two shafts shown. For each solution, determine if the value of Φ on the interior boundaries is sufficiently constant for an accurate solution. For the second cross-section, were the values of Φ the same at the two interior boundaries? Is this what you were expecting?

P4. *Seepage under a concrete dam.* A concrete gravity dam has a sheet pile located at its upstream face to help reduce the uplift pressure under the dam. For the conditions shown, determine the uplift pressure when the sheet pile is 14.0 feet in length. Soil A has a permeability of 0.2 ft/day, and soil B has a permeability of 0.7 ft/day. The upstream boundaries and downstream boundaries should be far removed from the region of analysis. Make two analyses, one with impervious boundaries both upstream and downstream, and one with constant heads. Explain any difference in the two solutions obtained, and discuss whether you think your boundaries are far enough removed to produce accurate approximations to the uplift pressure.

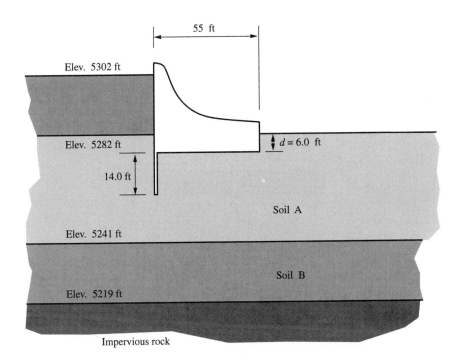

P5. *Flow of a viscous fluid in a heat exchanger.* Heat exchangers, such as the one shown below, are an integral part of chemical processing plants. One fluid is passed through the inner pipe while a second fluid flows between the inner pipe and the outer pipe. To enhance the heat exchange, fins are added to the inner pipe as shown on the right. The greater the number of fins, the greater the efficiency of the heat exchanger. However, as the number of fins increases, the pressure gradient necessary to create a given flow of the outside fluid also increases. In the design of the heat exchanger, it is necessary to know the velocity field of the outer fluid around the fins and the pressure gradient necessary to drive it.

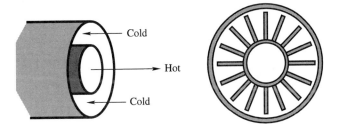

Use the formulation for parallel flow of a viscous fluid to determine the velocity field for a heat exchanger with 32 fins. The inner pipe has an outside diameter of 70 mm, the outer pipe has an inside diameter of 140 mm, the fin length is 30 mm, and the fin thickness is 1.5 mm. Determine your velocity in terms of a unit pressure gradient and a unit viscosity.

Work the problem using symmetry; hence, your mesh will be a wedge with one side the fin and the other side the plane of symmetry between two fins. Note also that a plane of symmetry exists between the end of the fin and inner wall of the outside pipe. This plane, however, is not in line with the edge of the fin.

Run a second analysis and assume the fin's thickness negligible so that its edge is along the line of symmetry. Is the error significant?

P6. *A capacitive probe.* Electrostatic capacitive probes are used to detect changes within dielectric materials without the need for physical intrusion into the object. The process can be used to detect cracks, voids, or other inconstancies in a specimen. In the medical profession, such techniques are used to detect tumors.

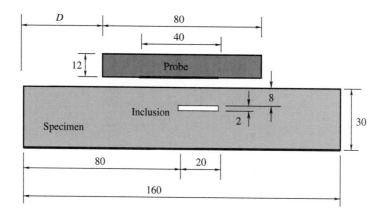

Shown here is a probe that will be used to detect possible voids in the specimen. A positive electric potential will be maintained along the dark band shown on the underside of the probe. A zero electric potential will be maintained on the bottom of the specimen along its entire length. The change in admittance (the ratio of current change to voltage) will then be measured as the probe is moved from left to right along the top of the specimen. From these measurements it will be possible to detect (or see) the location and size of any void.

Let the unit of length shown be L, and the positive potential on the underside of the probe be $1.0Q/L$. The dielectric material constants are

Air	1.0
Specimen	15.0
Probe	7.0

Assume the probe is lowered so that the space between the probe and specimen is $1.0L$. Conduct a finite element analysis to "see" the void by the deflection of equipotential lines in its vicinity. Conduct three analyses:

(a) Probe at $L = 40$, no void
(b) Probe at $L = 40$ with void
(c) Probe at $L = 25$ with void

Use a rectangular region for your analysis with the lower boundary corresponding to that of the specimen, the left and right sides corresponding to that of the specimen, and the upper boundary $10L$ units above the top of the probe. Consider all sides of the region of analysis as being no-flux boundaries.

CHAPTER **8**

HIGHER-ORDER ELEMENTS

In the previous chapter we used three-node elements that produced, within each element, a linear approximation of the primary variable. It is possible, and sometimes desirable, to use higher-order approximating functions. However, for these higher-order elements difficulties arise in maintaining continuity of the approximating function between elements. This problem has been solved through the use of isoparametric elements that are formed by mapping elementary-shaped elements into more complex configurations. The mapping is performed using the element shape functions themselves. Before presenting this solution, however, we will illustrate the problem. It is only through an appreciation of the problem that one can fully appreciate the solution.

8.1 THE PROBLEM: CONTINUITY

We want to create a finite element approximation of a function using quadrilateral elements over a domain similar to the one shown in Fig. 8.1. These elements will have four nodes, and within each element, the approximation will be defined in terms of the nodal Φ values, i.e.,

$$\Phi(x, y) = \left\lfloor N_1 \quad N_2 \quad N_3 \quad N_4 \right\rfloor \left\{ \begin{array}{c} \Phi_1 \\ \Phi_2 \\ \Phi_3 \\ \Phi_4 \end{array} \right\} \tag{8.1}$$

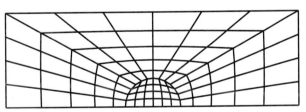

Figure 8.1. A domain divided into quadrilateral elements.

It is necessary that the above shape functions be defined so that when they are pieced together to form a global approximation, the global approximation is continuous across all element boundaries.

The first approach we used to define the shape functions for triangular elements was to write a polynomial for each element with as many independent parameters as there were element nodes. Hence, for our four-node element we select

$$\Phi(x, y) = \alpha_1 + \alpha_2 x + \alpha_3 y + \alpha_4 xy$$

$$\Phi(x, y) = \left\lfloor 1 \quad x \quad y \quad xy \right\rfloor \begin{Bmatrix} \alpha_1 \\ \alpha_2 \\ \alpha_3 \\ \alpha_4 \end{Bmatrix} = \lfloor M \rfloor \{\alpha\} \tag{8.1}$$

with the origin, say, at any one of the nodes or the centroid of the element. If we now continue along the same lines as in Chapter 5, we have

$$\Phi = \lfloor M \rfloor \{\alpha\}$$

$$\{\Phi\}_e = [A]\{\alpha\}$$

$$\{\alpha\} = [A]^{-1}\{\Phi\}_e \tag{8.3}$$

$$\Phi = \lfloor M \rfloor [A]^{-1}\{\Phi\}_e$$

$$\Phi = \lfloor N \rfloor \{\Phi\}_e$$

where $\lfloor N \rfloor$ is the array of our desired shape functions. Unfortunately, due to the xy term in the approximation we selected, this approach produces shape functions that vary parabolically along any side that is not parallel to either the x axis or the y axis. Therefore, even when adjacent elements have the same nodal values for Φ, the approximation might not be continuous across their common boundary. This is illustrated in Fig. 8.2.

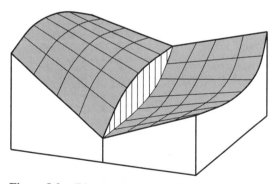

Figure 8.2.   Discontinuity across element boundaries.

Having met with failure using the first method, we now consider the second approach discussed in Chapter 5. That is, for a given node, we use the equations of the sides that are not contiguous to it as factors for its shape function. Thus, on these sides its value would be identically zero. This ensures that only the nodes contiguous to a given side determine the approximating function along that side. Because adjacent elements have common nodes, we might believe they would have identical approximating functions (or identical shape functions) along their common boundary. However, this is not the case. For example, consider the two elements, *A* and *B*, shown in Fig. 8.3.

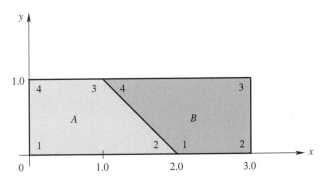

Figure 8.3.   Two quadrilateral elements.

If the equations for sides 4-1 and 1-2 of element *A* are used as factors for the shape function at node 3, we have

$$N_3(x, y) = xy \tag{8.4}$$

where *x* and *y* are the global coordinates shown in the figure. Note that N_3 is zero along sides 4-1 and 1-2, and its value at node 3 is unity.

Now consider the shape function for the same node in element *B*. If we follow the same technique to arrive at its equation, its factors would be the equations for sides 1-2 and 2-3 of this element. Thus, N_4 of element *B* becomes

$$N_4(x, y) = \tfrac{1}{2}(3.0 - x)(y) \tag{8.5}$$

Note again that this shape function is zero along sides to which it is not contiguous and is unity at its node. The values of the two shape functions along their common side are shown in Fig. 8.4. Clearly, if these shape functions were used for a finite element analysis, the approximating function would not be a continuous function within the domain of analysis.

In summary, in order that a piecewise FEM approximating function be continuous across element boundaries, it is necessary that the shape functions be uniquely defined along any element side by information from nodes contiguous to that side. Neither nodal values nor nodal locations associated with nodes not contiguous to that side must affect the approximation along that side. It was shown that the two approaches used in Chapter 5 failed to produce shape functions for the four-node element that satisfied this requirement. Of course, these methods were never proposed in the literature; they have been presented here only to demonstrate the difficulty in maintaining continuity when elements other than three-node triangles are used. We now show how this problem can be solved.

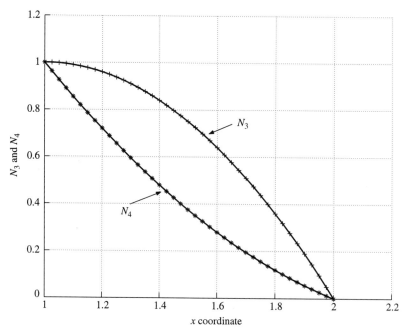

Figure 8.4. Shape function N_3 of element A and N_4 of element B.

8.2 THE SOLUTION: ISOPARAMETRIC ELEMENTS

The desired solution to our problem is found through the use of isoparametric elements. Not only will this approach allow us to develop shape functions for our four-node quadrilateral element, it is general enough to use for all higher-order elements introduced later in this chapter. It can be summarized as follows:

1. For a given higher-order element, restrict its geometry so that shape functions that meet the requirements necessary to maintain continuity between elements are easily defined.

2. Use these functions, which are continuous between elements, to map the simple element shapes into more complex element shapes.

3. Because the function defined in step 1 is continuous and the mapping used in step 2 is continuous, the final mapped finite element approximation over the more complex domain will be continuous.

In what follows, we first describe the element of restricted geometry (the parent element), then describe the mapping of this element into the more complex geometry, and then discuss how the two come together to give us our desired approximating function.

8.2.1 The Parent Element. To overcome the difficulty of maintaining continuity across element boundaries, we restrict the element to be rectangular with sides parallel to the coordinate axes as shown in Fig. 8.5. We have labeled the axes u and v rather than x and y because we will eventually map this element onto the x-y plane. Also, we chose the u-v origin to be at the center of the element, with the element boundaries at ± 1. This choice was made to correspond to Gaussian coordinates, which we will use to perform our element integrations. This is similar to the approach we used in Chapter 4. The shape functions for this element, in terms of the axes shown, can be written by inspection if we follow the procedure of using the equations of the sides opposite the node as factors. Hence,

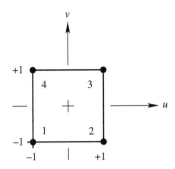

Figure 8.5. Rectangular four-node element.

$$N_1(u, v) = (1/4)(1 - u)(1 - v)$$

$$N_2(u, v) = (1/4)(1 + u)(1 - v)$$

$$N_3(u, v) = (1/4)(1 + u)(1 + v)$$ (8.6)

$$N_4(u, v) = (1/4)(1 - u)(1 + v)$$

Notice that these functions vary linearly along each side of the element. This is a consequence of restricting the sides to be parallel with the coordinate axes. Hence, along any side, either u or v is constant and the uv term in Eqs. 8.6 is linear. Note, however, that interior to the element, along lines not parallel to one of the coordinate axes, the shape functions will not vary linearly.

Because the shape functions are linear along all four sides of the element, we are assured that they will provide a continuous approximation if they are pieced together with similar elements as shown in Fig. 8.6.

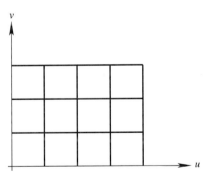

Figure 8.6. A mesh of rectangular elements.

Unfortunately, a mesh of rectangular elements is very limited. However, if our approximation is continuous over this simple rectangular region, and we map it onto another region in the x-y plane using a continuous mapping function, then the approximation will be continuous in the x-y plane even though the mapped elements may no longer be rectangular. Because our piecewise approximating function is itself continuous, we can use it as our mapping function. Thus, we will use the same parametric equations to describe our mapping as we do our approximation—hence, the term *isoparametric elements*.

8.2.2 The Mapping. We now show how to use the element shape functions as mapping functions. Consider the four shape functions defined by Eq. 8.6. These shape functions can be used to define any variable within the element shown in Fig. 8.5. Thus, if we associate every point in the u-v plane with a corresponding point in the x-y plane, that association can be represented using the the four shape functions. That is,

$$x(u, v) = \lfloor N(u, v) \rfloor \{X\}$$

$$y(u, v) = \lfloor N(u, v) \rfloor \{Y\}$$

(8.7)

where X and Y are the coordinates that the four nodes will have in the x-y plane. For example, if we wish an element to be mapped onto the x-y plane in the position shown in Fig. 8.7, we see that the four nodal coordinates would have to be

$$(x_1, y_1) = (5, 2)$$
$$(x_2, y_2) = (10, 3)$$
$$(x_3, y_3) = (7, 11)$$
$$(x_4, y_4) = (1, 4)$$

(8.8)

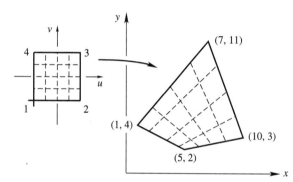

Figure 8.7. Mapping of a parent element to the x-y plane.

For this particular mapping, we would have

$$x(u, v) = \lfloor N \rfloor \{X\} = \lfloor N_1 \quad N_2 \quad N_3 \quad N_4 \rfloor \begin{Bmatrix} 5 \\ 10 \\ 7 \\ 4 \end{Bmatrix}$$

(8.9)

$$y(u, v) = \lfloor N \rfloor \{Y\} = \lfloor N_1 \quad N_2 \quad N_3 \quad N_4 \rfloor \begin{Bmatrix} 2 \\ 3 \\ 11 \\ 4 \end{Bmatrix}$$

This assures us that points in the *u-v* plane get mapped to points in the *x-y* plane in such a manner that the four nodal points will have the coordinates prescribed by {*X*} and {*Y*}. Also illustrated in Fig. 8.7 is the fact that all lines parallel to the *u* and *v* axes are mapped as straight lines in the *x-y* plane, not just the sides of the element. The use of the shape functions to perform the mapping guarantees that the elements spanning the *x-y* domain will be continuous. This is illustrated in Fig. 8.8.

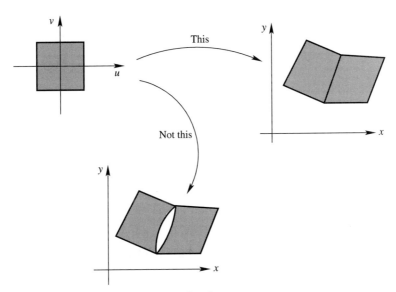

Figure 8.8.   Continuous mapping for element geometry.

8.3 PUTTING IT TOGETHER

Previously, we defined each side of an element's boundary by using the coordinates of the nodes on that particular side, and only those nodal coordinates. We also defined the approximating function along each side by using nodal values at the nodes on that particular side, and only those values. Therefore, any two elements that use the same nodes to define a side will have identical boundaries everywhere along that side and will have identical approximating functions along that side, regardless of the location of any of the other nodes and the values of the approximating function at these nodes. Hence, there will be no discontinuity across these boundaries under any condition.

This continuity can be easily visualized for our four-node element. The element sides will always be straight lines and the approximating function, Φ, will always vary linearly along each side as shown in Fig. 8.9. Therefore, elements that share a side will have identical approximating functions and will "fit" together to form a continuous surface across their common boundary.

The same holds true for higher-order elements (which we will soon consider). For these elements there will be more than two nodes associated with a side, and these nodes will define a curve in the *x-y* plane. However, we will show that for any given side, the coordinates of all points along the side and the corresponding Φ value are uniquely defined by their nodal point values along the side and no other values. Therefore, any two elements that have common nodal points along a common side will have identical *x*, *y*, and Φ values at every point along that side. Therefore, continuity of the geometry and the approximating function along these boundaries will be maintained, no matter what.

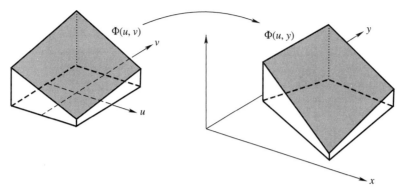

Figure 8.9.   Mapping of Φ values.

8.4 CREATING THE K MATRIX

In the previous sections we showed that continuous, piecewise approximating functions can be obtained by defining them in the u-v domain and then mapping them into the x-y domain. In a finite element analysis, this method is actually used in reverse. That is, given elements in the x-y plane, we determine their u-v counterparts and from there determine the element stiffness matrices and the right-hand side vectors. Because these quantities are defined with respect to x-y coordinates, this amounts to a change in variables from these coordinates to u-v coordinates. This will require integration and differentiation with respect to x-y coordinates within the u-v domain. We now show how this is done within a finite element code.

8.4.1 Integration in the u-v Plane.   We demonstrate the technique for integrating a function with the simple example of finding the volume under the Φ surface as is illustrated in Fig. 8.10. That is, we seek

$$V = \lim_{n \to \infty} \sum_{i=1}^{n} \Phi(x_i, y_i) \Delta A = \int_A \Phi \, dA \tag{8.10}$$

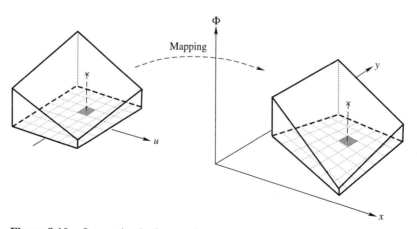

Figure 8.10.   Integration in the u-v plane.

Here we show that the integral is the limit of the summation shown. This understanding is necessary for an understanding of what follows.

To perform the integration, we consider the parent element in the u-v plane divided into differential areas as shown. The corresponding areas are shown in the x-y plane. The procedure is to evaluate Φ at some point (say the centroid) in each differential area of the u-v plane and multiply this value by the corresponding dA in the x-y plane. When the entire area in the u-v plane has been covered by this procedure, the entire area in the x-y plane will have been covered, thus completing the integration. We are able to do this because the value of Φ at any point in the u-v plane is the same as its value at the mapped point in the x-y plane. The critical step in this procedure is to determine the ΔA in the x-y plane from the corresponding area in the u-v plane.

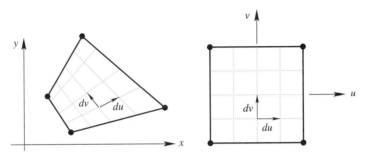

Figure 8.11. Mapping of a differential area.

Consider the four-node element in the x-y plane as shown in Fig. 8.11. We now understand that it can be considered as an element mapped from the unit square element in the u-v plane. Next, consider the differentials du and dv as two vectors in the u-v plane that are mapped into the x-y plane as shown in the figure. The differential area is then written as

$$dA = |d\vec{u} \times d\vec{v}| \qquad (8.11)$$

To evaluate this cross-product we must first determine the x and y components of each vector. The differential du represents the displacement from the point (u, v) to the point $(u + du, v)$. The corresponding displacement in the x-y plane would be from the point (x, y) to the point $(x + dx, y + dy)$. Note that a differential change du in the u-v plane causes a differential change in the x-y plane equal to

$$dx = \frac{\partial x}{\partial u} du$$
$$dy = \frac{\partial y}{\partial u} du \qquad (8.12)$$

Hence, these are the x and y components of the transformed vector du. We write, therefore,

$$d\vec{u} = \frac{\partial x}{\partial u} du \, \hat{i} + \frac{\partial y}{\partial u} du \, \hat{j} \qquad (8.13)$$

and likewise,

$$d\vec{v} = \frac{\partial x}{\partial v} dv \, \hat{i} + \frac{\partial y}{\partial v} dv \, \hat{j} \qquad (8.14)$$

where $\hat{i}$ and $\hat{j}$ are unit vectors in the x and y directions. The differential area can now be written as

$$dA = \left(\frac{\partial x}{\partial u} \frac{\partial y}{\partial v} - \frac{\partial y}{\partial u} \frac{\partial x}{\partial v} \right) du\, dv \tag{8.15}$$

or

$$dA = \left\| \begin{matrix} \dfrac{\partial x}{\partial u} & \dfrac{\partial x}{\partial v} \\[2mm] \dfrac{\partial y}{\partial u} & \dfrac{\partial y}{\partial v} \end{matrix} \right\| du\, dv = \|J\|\, du\, dv \tag{8.16}$$

where the 2×2 matrix is the Jacobian matrix and $\|J\|$ is its determinant. The integral can now be evaluated as

$$\int_A \Phi\, dA = \int_{u=-1}^{1} \int_{v=-1}^{1} \Phi(u, v) \|J\|\, du\, dv \tag{8.17}$$

Thus, we are able to evaluate our integral defined in the x-y plane by performing our integration in the u-v plane. For our finite element analysis, the Jacobian matrix and its determinate are evaluated using

$$\frac{\partial x}{\partial u} = \lfloor \partial N/\partial u \rfloor \{ X \}$$

$$\frac{\partial x}{\partial v} = \lfloor \partial N/\partial v \rfloor \{ X \}$$

$$\frac{\partial y}{\partial u} = \lfloor \partial N/\partial u \rfloor \{ Y \} \tag{8.18}$$

$$\frac{\partial y}{\partial v} = \lfloor \partial N/\partial v \rfloor \{ Y \}$$

8.4.2 Differentiation in the u-v Plane. We have seen how $\Phi(x, y)$ may be integrated in the u-v plane. However, the integrand of the K matrix,

$$[N']^T [R][N']$$

contains derivatives of the shape functions with respect to x and y that will have to be evaluated in the u-v plane. Because our shape functions are written in terms of u and v, the chain rule is used to obtain

$$\frac{\partial N_i}{\partial x} = \frac{\partial N_i}{\partial u} \frac{\partial u}{\partial x} + \frac{\partial N_i}{\partial v} \frac{\partial v}{\partial x} \tag{8.19}$$

and

$$\frac{\partial N_i}{\partial y} = \frac{\partial N_i}{\partial u} \frac{\partial u}{\partial y} + \frac{\partial N_i}{\partial v} \frac{\partial v}{\partial y} \tag{8.20}$$

or, in matrix notation,

$$
\left\{ \begin{array}{c} \dfrac{\partial N_i}{\partial x} \\[2mm] \dfrac{\partial N_i}{\partial y} \end{array} \right\} = \left[\begin{array}{cc} \dfrac{\partial u}{\partial x} & \dfrac{\partial v}{\partial x} \\[2mm] \dfrac{\partial u}{\partial y} & \dfrac{\partial v}{\partial y} \end{array} \right] \left\{ \begin{array}{c} \dfrac{\partial N_i}{\partial u} \\[2mm] \dfrac{\partial N_i}{\partial v} \end{array} \right\} \tag{8.21}
$$

We now use the somewhat standard notation

$$
\left[J \dfrac{(u,\,v)}{(x,\,y)} \right] = \left[\begin{array}{cc} \dfrac{\partial u}{\partial x} & \dfrac{\partial u}{\partial y} \\[2mm] \dfrac{\partial v}{\partial x} & \dfrac{\partial v}{\partial y} \end{array} \right] \tag{8.22}
$$

and

$$
\left[J \dfrac{(x,\,y)}{(u,\,v)} \right] = \left[\begin{array}{cc} \dfrac{\partial x}{\partial u} & \dfrac{\partial x}{\partial v} \\[2mm] \dfrac{\partial y}{\partial u} & \dfrac{\partial y}{\partial v} \end{array} \right] \tag{8.23}
$$

Note that

$$
\left\{ \begin{array}{c} \dfrac{\partial N_i}{\partial x} \\[2mm] \dfrac{\partial N_i}{\partial y} \end{array} \right\} = \left[J^T \dfrac{(u,\,v)}{(x,\,y)} \right] \left\{ \begin{array}{c} \dfrac{\partial N_i}{\partial u} \\[2mm] \dfrac{\partial N_i}{\partial v} \end{array} \right\} \tag{8.24}
$$

where

$$
\left[J^T \dfrac{(u,\,v)}{(x,\,y)} \right] = \left[J \dfrac{(u,\,v)}{(x,\,y)} \right]^T \tag{8.25}
$$

The differentials transform as

$$
\left\{ \begin{array}{c} dx \\ dy \end{array} \right\} = \left[J \dfrac{(x,\,y)}{(u,\,v)} \right] \left\{ \begin{array}{c} du \\ dv \end{array} \right\} \tag{8.26}
$$

and

$$
\left\{ \begin{array}{c} du \\ dv \end{array} \right\} = \left[J \dfrac{(u,\,v)}{(x,\,y)} \right] \left\{ \begin{array}{c} dx \\ dy \end{array} \right\} \tag{8.27}
$$

Hence,

$$\left[J \frac{(u, v)}{(x, y)} \right] = \left[J \frac{(x, y)}{(u, v)} \right]^{-1} \tag{8.28}$$

The last equation is important since it allows us to write

$$\left\{ \begin{array}{c} \dfrac{\partial N_i}{\partial x} \\[2mm] \dfrac{\partial N_i}{\partial y} \end{array} \right\} = \left[J^T \frac{(x, y)}{(u, v)} \right]^{-1} \left\{ \begin{array}{c} \dfrac{\partial N_i}{\partial u} \\[2mm] \dfrac{\partial N_i}{\partial v} \end{array} \right\} \tag{8.29}$$

We recall that

$$dA = \det \left[J \frac{(x, y)}{(u, v)} \right] du\, dv$$

$$= \|J\| \, du\, dv \tag{8.30}$$

With the transformations

$$x(u, v) = [N(u, v)]\{X\}$$

$$y(u, v) = [N(u, v)]\{Y\} \tag{8.31}$$

we have

$$\frac{\partial x}{\partial u} = \lfloor \partial N/\partial u \rfloor \{X\} \qquad \frac{\partial y}{\partial u} = \lfloor \partial N/\partial u \rfloor \{Y\}$$

$$\frac{\partial x}{\partial v} = \lfloor \partial N/\partial v \rfloor \{X\} \qquad \frac{\partial y}{\partial v} = \lfloor \partial N/\partial v \rfloor \{Y\} \tag{8.32}$$

We can now obtain

$$[N'] = \begin{bmatrix} \dfrac{\partial N_1}{\partial x} & \dfrac{\partial N_2}{\partial x} & \cdots & \dfrac{\partial N_n}{\partial x} \\[3mm] \dfrac{\partial N_1}{\partial y} & \dfrac{\partial N_2}{\partial y} & \cdots & \dfrac{\partial N_n}{\partial y} \end{bmatrix} \tag{8.33}$$

where n = number of nodes per element, using

$$\begin{bmatrix} \dfrac{\partial N_1}{\partial x} & \cdots & \dfrac{\partial N_n}{\partial x} \\[3mm] \dfrac{\partial N_1}{\partial y} & \cdots & \dfrac{\partial N_n}{\partial y} \end{bmatrix} = \begin{bmatrix} \dfrac{\partial u}{\partial x} & \dfrac{\partial v}{\partial x} \\[3mm] \dfrac{\partial u}{\partial y} & \dfrac{\partial v}{\partial y} \end{bmatrix} \begin{bmatrix} \dfrac{\partial N_1}{\partial u} & \cdots & \dfrac{\partial N_n}{\partial u} \\[3mm] \dfrac{\partial N_1}{\partial v} & \cdots & \dfrac{\partial N_n}{\partial v} \end{bmatrix} \tag{8.34}$$

which we write as

$$[N'(x, y)] = \left[J^T \frac{(u, v)}{(x, y)} \right] [N'(u, v)] \tag{8.35}$$

The element K matrix

$$[K]_e = \int_{V_e} [N'(x, y)]^T [R] [N'(x, y)] \, dV \tag{8.36}$$

can now be written as

$$[K]_e = \int_{V_e} [N'(u, v)]^T \left[J \frac{(x, y)}{(u, v)} \right]^{-1} [R] \left[J^T \frac{(x, y)}{(u, v)} \right]^{-1} [N'(u, v)] \|J\| \, du \, dv \tag{8.37}$$

where, in writing the first Jacobian matrix, we have used

$$\left[\left[J^T \frac{(x, y)}{(u, v)} \right]^{-1} \right]^T = \left[J \frac{(x, y)}{(u, v)} \right]^{-1} \tag{8.38}$$

We now have, in more compact notation,

$$[K]_e = \int_{V_e} [N']^T [J]^{-1} [R] [J^T]^{-1} [N'] \|J\| \, du \, dv \tag{8.39}$$

8.4.3 Surface Integrals. We have shown how the finite element volume integrals can be integrated in the u-v plane. We must now consider the corresponding integration of the surface integrals that appear in the weak form of governing equations. Let

$$\int_S \delta\Phi(S) q(S) \, dS \tag{8.40}$$

represent our surface integral, where we have emphasized that the weighting function (represented by the independent variation of the variable Φ) and the surface flux, q, are evaluated on the surface over which the integration takes place. If this integration is to be carried out in the u-v plane, it will be along one of the sides of the parent element. For our four-node parent element, as well as for parent elements of higher order, this will be a straight line. However, for elements with more nodes than four, the sides are not necessarily straight lines in the x-y plane. Let

$$S = S(u) \tag{8.41}$$

represent the mapping along a straight side of the parent element to the corresponding side in the x-y plane as shown in Fig. 8.12.

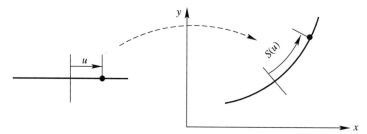

Figure 8.12. One-dimensional mapping for surface quadrature.

To evaluate our integral along u, we convert dS to du and write our integrand in terms of u. Thus,

$$dS = \frac{dS}{du} du \tag{8.42}$$

and

$$\int_u \delta\Phi(u) q(u) \frac{dS}{du} du \tag{8.43}$$

The mapping of u to S will be performed by mapping u to the x and y coordinates of S. Therefore, we use the relationship

$$dS = \left(dx^2 + dy^2 \right)^{1/2} \tag{8.44}$$

where

$$dx = \left(\frac{dx}{du} \right) du$$

$$dy = \left(\frac{dy}{du} \right) du \tag{8.45}$$

and write

$$dS = \left[\left(\frac{dx}{du} \right)^2 + \left(\frac{dy}{du} \right)^2 \right]^{1/2} du \tag{8.46}$$

Because of the similarity of the above expression to the mapping of differential areas, we use the following notation:

$$\|J_s\| = \frac{dS}{du} = \left[\left(\frac{dx}{du} \right)^2 + \left(\frac{dy}{du} \right)^2 \right]^{1/2} \tag{8.47}$$

and write our integral as

$$\int_u \delta\Phi(u)\, q(u)\, \|J_s\|\, du \tag{8.48}$$

The mapping used for a side of an element is, of course, the same mapping that is used for the entire element. Thus, the shape functions are again used, but only their values along the side that will be mapped. The shape functions, therefore, reduce to one-dimensional shape functions as shown in Fig. 8.13.

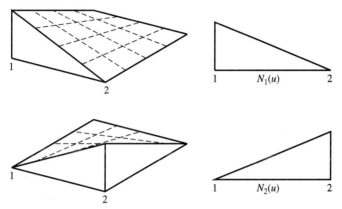

Figure 8.13.   The $N(u)$ shape functions used for surface integration.

For this element, there are only two nodes per side, so our mapping is

$$x(u) = \lfloor N_1(u) \quad N_2(u) \rfloor \begin{Bmatrix} x_1 \\ x_2 \end{Bmatrix}$$

$$\tag{8.49}$$

$$y(u) = \lfloor N_1(u) \quad N_2(u) \rfloor \begin{Bmatrix} y_1 \\ y_2 \end{Bmatrix}$$

This allows us to calculate $\|J_s\|$ using

$$\left.\begin{aligned} \frac{dx}{du} &= \lfloor dN/du \rfloor \{x\} \\[2mm] \frac{dy}{du} &= \lfloor dN/du \rfloor \{y\} \end{aligned}\right\} \quad \text{for } x \text{ and } y \text{ along } S \tag{8.50}$$

In addition, the variation of Φ along the side is given by

$$\delta\Phi = \lfloor N(u) \rfloor \{\delta\Phi\} \tag{8.51}$$

Thus,

$$\int_S \delta\Phi\, q\, dS = \lfloor \delta\Phi \rfloor \int_u \{ N(u) \}\, q(u) \|J_s\|\, du \tag{8.52}$$

Finally, it is often convenient to use nodal point values to define surface flux. Thus, $q(u)$ is determined using

$$q(u) = \lfloor N(u) \rfloor \{ q \} \tag{8.53}$$

8.5 ADDITIONAL ELEMENTS

We now introduce some additional elements that are commonly used in the approximation of solutions to second-order partial differential equations in two dimensions. These elements are defined in the u-v plane and are to be considered as parent elements for more general isoparametric elements obtained by mapping them onto an x-y plane.

8.5.1 The Six-Node Triangular Element. The six-node triangular element is popular because it provides a complete second-degree polynomial in two dimensions for the approximating function; that is, all terms of the general second-degree polynomial

$$P = A + Bu + Cv + Du^2 + Euv + Fv^2$$

are present in the collection of its six shape functions. Figure 8.14 defines the geometry of the triangle in the u-v plane. The coordinates shown are those most often used for Gaussian quadrature, although they are usually

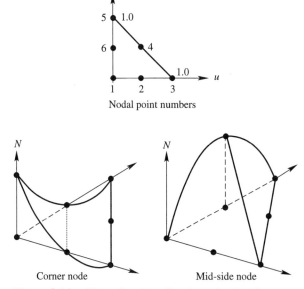

Figure 8.14.   Shape functions for six-node triangle.

interpreted as area coordinates rather than Cartesian coordinates as shown. The six shape functions associated with each node shown must

1. Be equal to unity at the node they represent and zero at all other nodes
2. Be identically zero along any noncontiguous side
3. Vary parabolically along any contiguous side

The last two requirements result in the approximating function being completely defined along any side by its values at the nodes contiguous to that side. Thus, it will be continuous across the interface of two adjoining elements that share the same nodes along their common boundary.

With this understanding, the shape functions are easily obtained. Consider any one of the three vertices, for example, node 1. Its shape function must be zero along the line 3-4-5 and at nodes 2 and 6. If we take as factors for $N_5(u, v)$ the equation of line 3-4-5 and the equation of line 2-6, then we have such a function. These are the two factors shown for N_5 in the following equation. We must also make sure that its value is equal to unity at node 1, which it clearly is. Finally, we want to check that it varies parabolically along sides 1-2-3 and 5-6-1 of the element. Again, this is easily verified.

Now consider any one of the mid-side points, for example, node 2. It will have to have as factors the equation for line 3-4-5 and the equation for line 5-6-1. These are the two noncontiguous sides for this node. The two factors shown in the following equation for N_6 are these equations. Note that these sides include all other nodes; there are no remaining nodes we must be concerned about in the requirement that the shape function be zero at all other nodes. Next we make N_2 equal to unity at node 2 by multiplying by -4. Finally, we want to make sure N_6 varies parabolically along its contiguous side (side 1-2-3), which it does.

The remaining shape functions can be determined in exactly the same manner. The entire set is

$$N_1 = (u + v - 1)(2u + 2v - 1) \tag{8.54}$$

$$N_2 = -4(u)(u + v - 1) \tag{8.55}$$

$$N_3 = (2u - 1)(u) \tag{8.56}$$

$$N_4 = 4(u)(v) \tag{8.57}$$

$$N_5 = (2v - 1)(v) \tag{8.58}$$

$$N_6 = -4(u + v - 1)(v) \tag{8.59}$$

These six functions have only two basic shapes as shown in Fig. 8.14: one shape that is common to all corner nodes and a second shape that is common to all mid-side nodes. Finally, you should visualize how this parent element might be mapped onto the x-y plane. The three nodes per side allow both the x and y coordinates along any given side to be mapped parabolically. Therefore, six-node triangular elements in the x-y plane can have parabolic sides.

8.5.2 The Eight-Node Quadrilateral Element. Our next element is the eight-node quadrilateral element shown in Fig. 8.15. Because each side of the quadrilateral has three nodes, the requirements placed on the eight shape functions are identical to those for the six-node triangle:

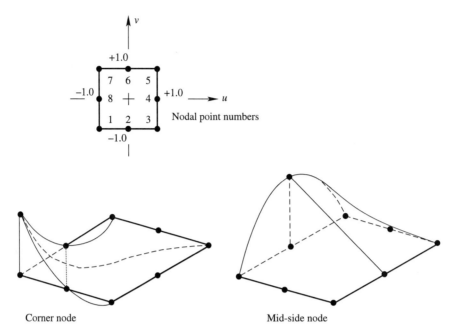

Figure 8.15.   Shape functions for eight-node quadrilateral.

1. They are equal to unity at the node they represent and zero at all other nodes.
2. They are identically zero along any noncontiguous side.
3. They vary parabolically along any contiguous side.

The eight shape functions, obtained in the same manner as described for the six-node element, are

$$N_1 = -(1/4)(1 - u)(1 - v)(1 + u + v)$$

$$N_2 = (1/2)(1 + u)(1 - u)(1 - v)$$

$$N_3 = -(1/4)(1 + u)(1 - v)(1 - u + v)$$

$$N_4 = (1/2)(1 + u)(1 - v)(1 + v)$$

$$N_5 = -(1/4)(1 + u)(1 + v)(1 - u - v) \qquad (8.60)$$

$$N_6 = (1/2)(1 + u)(1 - u)(1 + v)$$

$$N_7 = -(1/4)(1 - u)(1 + v)(1 + u - v)$$

$$N_8 = (1/2)(1 - u)(1 - v)(1 + v)$$

8.5.3 The Nine-Node Quadrilateral Element. For the elements thus far considered, all nodes were on the element's boundaries, and their shape functions were derived using linear equations that passed through two or more of these nodes. There is another class of elements, which are quadrilaterals with regularly spaced nodes that appear internal to the element as well as on its boundaries. Lagrangian interpolation polynomials are used to derive the shape functions for such elements; hence, they are referred to as Lagrangian elements.

A Lagrangian polynomial in one dimension is an nth-degree polynomial that has roots specified at n points and has a value of unity specified at one other point. Consider Fig. 8.16, which shows a fourth-degree polynomial having the roots (x_1, x_3, x_4, x_5) and having a value of unity at x_2. This polynomial can be written as

$$P_2(x) = \frac{(x - x_1)(x - x_3)(x - x_4)(x - x_5)}{(x_2 - x_1)(x_2 - x_3)(x_2 - x_4)(x_2 - x_5)} \tag{8.61}$$

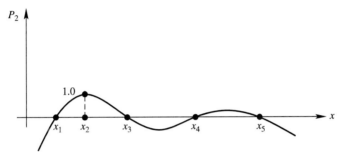

Figure 8.16. A Lagrangian polynomial.

From the numerator, it is clear that P_2 is zero at all the desired points. It is also clear that at $x = x_2$, the numerator is equal to the denominator; hence, $P_2 = 1.0$. The general form for the Lagrangian polynomial that we use is

$$P_k(x) = \frac{\Pi(x - x_j)}{\Pi(x_k - x_j)} \quad j = 1, n \quad (j \neq k) \tag{8.62}$$

where n is the total number of points including the point where P is equal to unity $(x = x_k)$.

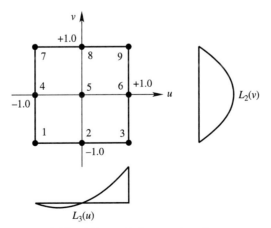

Figure 8.17. Nine-node Lagrangian element.

Because our shape functions are to equal zero at all nodes except the one they represent, Lagrangian polynomials are easily used for their development. As an example, let us consider the shape function for node 6 in the nine-node element shown in Fig. 8.17. Let $L_k(u)$ and $L_k(v)$ represent second-degree Lagrangian polynomials in u and v, respectively. Then

$$N_6 = L_3(u) \times L_2(v) \tag{8.63}$$

where $L_3(u)$ represents the polynomial in u associated with the third of its three nodes, and $L_2(v)$ represents the polynomial in v associated with the second of its three nodes. These are illustrated in the figure. Clearly, because $L_3(u)$ is a factor of N_6, it will be zero at nodes 1, 2, 4, 5, 7, and 8. Likewise, because $L_2(v)$ is also a factor, N_6 will be zero at nodes 1, 2, 3, 7, 8, and 9. However, both $L_3(u)$ and $L_2(v)$ are equal to unity at node 6; hence, our shape function itself will be equal to unity at its own node. In this same manner, all nine shape functions are easily written in terms of the three Lagrangian polynomials associated with either the three u coordinates or the three v coordinates represented in this element.

8.5.4 Surface Shape Functions. For the elements considered in this section, the number of nodes on a side equals 3. This means that there will be three surface shape functions $N(u)$ that vary parabolically along any given side. Figure 8.18 illustrates these shape functions for the eight-node quadrilateral. If the coordinate for u is at the center node and the length of the side is 2, then the equations for these functions are

$$N_1(u) = -\left(\tfrac{1}{2}\right)u(1 - u) \tag{8.64}$$

$$N_2(u) = \left(\tfrac{1}{2}\right)(1 + u)(1 - u) \tag{8.65}$$

$$N_3(u) = \left(\tfrac{1}{2}\right)(1 + u)u \tag{8.66}$$

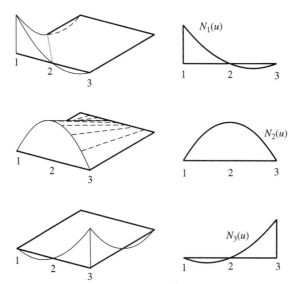

Figure 8.18. Surface shape functions for elements with three nodes per side.

8.6 DIMENSIONS OF SHAPE FUNCTIONS

For each of the parent elements given above, the u and v coordinates are dimensionless; that is, they are the ratio of the distance to a characteristic length associated with the element. For all of the elements we have considered, this characteristic length is unity. For this reason, the nondimensionality of these coordinates is obscure. Had we chosen the sides of our parent elements to range from 0 to $+a$, or $-a$ to $+a$, rather than 0 to $+1$ or -1 to $+1$, then our coordinates would have been u/a and v/a, rather than simply u and v. Because the coordinates are dimensionless, the shape functions themselves are dimensionless. Thus, when they are multiplied by nodal point values, the product has the dimensions associated with the nodal point quantity.

8.7 FINAL REMARKS

We have shown several elements and hinted that there are many more. It is natural to ask whether there is a "best" element. The answer is "Probably, but it would be difficult to determine and likely not worth the effort." However, there are some guidelines that spell out certain advantages and disadvantages. Here are a few things to think about.

The higher the order, the better the approximation within an element and over the domain of analysis. Hence, the rate of convergence increases with higher-order elements. Note, however, that this rate is measured in terms of the number of nodes, not computer time. More computer time is necessary to create the stiffness matrices for higher-order elements; in addition, higher-order elements create larger bandwidths, which in turn require a greater amount of computer time to solve the FEM equations.

One particular problem that arises with the use of low-order elements is associated with internal constraints imposed on the dependent variable. The most common occurrence of this is in the analysis of incompressible fluid flow. Here the requirement of incompressibility places constraints on the velocity field (i.e., constraints on the nodal point values of velocity). To understand the problem, you should realize that any approximation of a function can be understood as a constraint. Thus, finite element approximations constrain solutions to be in the class of functions to which the approximations belong. If in addition to this constraint, the problem also imposes constraints, there simply may be too many constraints to obtain a decent solution. In these cases, it is necessary to use higher-order elements.

In conclusion,

- For two-dimensional problems described by second-order partial differential equations with no internal constraints, most practitioners use one of the elements already described (i.e., those with three, four, six, eight, or nine nodes).

- If any element does have a clear advantage, it will be due to the specific problem being analyzed more than anything else. In these cases, the advantage will be one that the user will clearly understand, and the best (or near best) choice will be rather transparent. Users usually arrive at their own favorite element.

- If a problem has internal constraints, special care should be taken to make sure the approximation is not overconstrained. Literature on the analysis of incompressible materials is abundant and should be studied before a user enters this area of analysis.

- It is possible to use more than one type of element within a single mesh. It is necessary only that continuity of the approximating function be maintained across element boundaries. This usually means that the two connecting elements have the same number of nodes on their connecting side.

- Elements for three-dimensional analysis are easily obtained by a simple extension of the methods thus far discussed.

- The elements presented cannot be used in the analysis of higher-order partial differential equations. The weak forms of these equations demand smoother functions than our current elements provide. In particular, they require continuity of the approximation function and its first derivatives. An introduction to the use of the finite element method for the solution of higher-order differential equations is given in Chapter 12.

EXERCISES

Study Problems

S1. Given a four-node rectangular element with the following values,

Node	XORD	YORD	PHI
1	5.0	3.0	20
2	14.0	5.0	26
3	9.0	13.0	36
4	3.0	8.0	32

    (a) Using the shape functions to define the transformation function, determine the x-y coordinates of the point corresponding to the centroid of the parent element.
    (b) Determine the value of Φ at this same point.
    (c) Determine the Jacobian (matrix) of the transformation at the centroid of the parent element.
    (d) Determine the derivative of Φ with respect to x at the the point corresponding to the centroid of the parent element.

S2. Verify Eq. 8.38.

Numerical Experiments and Code Development

N1. Write a MATLAB function that gives the value of the shape function corresponding to node N at coordinates (u, v) for
    (a) A four-node quadrilateral element
    (b) A six-node triangular element
    (c) An eight-node quadrilateral element
    (d) A nine-node quadrilateral element

N2. Write a function that returns the values of dN/du and dN/dv for the shape function corresponding to node N at coordinates (u, v) for
    (a) A four-node quadrilateral element
    (b) A six-node triangular element
    (c) An eight-node quadrilateral element
    (d) A nine-node quadrilateral element

N3. Write a computer code that plots a Lagrangian polynomial for a given set of n points and a given data set of nodal values.

N4. Write a computer code that plots the derivative of a Lagrangian polynomial for a given set of n nodes and a given set of nodal values.

N5. Write a computer code that will plot all shape functions for
   (a) A four-node quadrilateral element
   (b) A six-node triangle element
   (c) An eight-node quadrilateral element
   (d) A nine-node quadrilateral element

A FEM PROGRAM FOR TWO-DIMENSIONAL BOUNDARY VALUE PROBLEMS

We are now ready to consider how to incorporate higher-order elements into a computer code. We demonstrate this through developing a finite element program for the analysis of second-order, two-dimensional, linear partial differential equations used to describe steady-state processes, that is, those processes that are in equilibrium and are void of any transient characteristics.

9.1 GOVERNING EQUATION

The strong form of our governing equation is

Within V:

$$\frac{\partial}{\partial x}\left[R_x\frac{\partial \Phi}{\partial x}\right] + \frac{\partial}{\partial y}\left[R_y\frac{\partial \Phi}{\partial y}\right] + B_x\frac{\partial \Phi}{\partial x} + B_y\frac{\partial \Phi}{\partial y} + G\Phi + H = 0.0 \qquad (9.1)$$

On S:

$$R_x\frac{\partial \Phi}{\partial x}n_x + R_y\frac{\partial \Phi}{\partial y}n_y = q^* \qquad (9.2)$$

$$\Phi = \Phi^* \qquad (9.3)$$

The equation can be written in other forms. In particular, the terms B_x and B_y could be shown inside the first-derivative operators rather than outside. When these terms are constant, there is no difference; however, we will use this equation to describe applications for which these terms are not constant. The particular form used depends on the physical law the equation describes. For many applications, the equation is used to describe the conservation of a quantity that is a function of Φ. In such applications, the first-derivative terms represent the convection of this quantity due to flow of an incompressible fluid. When this is the case, there is again no difference between the two forms of the equation. However, it is important to be aware of the different forms. Note that it is always possible to write one form in terms of another form. For example, had we incorporated B_x and B_y in the operators, then

$$\frac{\partial}{\partial x}\{B_x\Phi\} + \frac{\partial}{\partial y}\{B_y\Phi\} = \frac{\partial B_x}{\partial x}\Phi + \frac{\partial B_y}{\partial y}\Phi + B_x\frac{\partial \Phi}{\partial x} + B_y\frac{\partial \Phi}{\partial y} \qquad (9.4)$$

and the first terms on the right-hand side could be incorporated into the definition of G in Eq. 9.1.

It is necessary that R_x and R_y be of equal sign throughout the region of analysis. If this is not the case, the governing equation is not elliptic and the problem is no longer a boundary value problem. Although the sign they both have can be either negative or positive, the physical significance for most problems is best understood when it is positive.

It is also necessary to specify (and hence know) one or the other of the boundary conditions at all points on S. In what follows, we continue our practice of not stating which condition is specified. Rather, we interpret our equations as true statements of the physical process, not depending on knowledge of either the solution or of any specific boundary condition. Eventually, however, we will have to specify one or the other boundary condition to obtain a unique solution to a specific problem.

9.1.1 First Weak Form. There are two different but useful weak forms of our problem. We begin by deriving the one most often used and the one used in the finite element code given in this chapter. From the strong form, we know

$$\int_V \delta\Phi \left\{ \frac{\partial}{\partial x}\left[R_x \frac{\partial \Phi}{\partial x}\right] + \frac{\partial}{\partial y}\left[R_y \frac{\partial \Phi}{\partial y}\right] + B_x \frac{\partial \Phi}{\partial x} + B_y \frac{\partial \Phi}{\partial y} + G\Phi + H \right\} dV = 0 \tag{9.5}$$

is true for all $\delta\Phi$. Integration by parts gives

$$\int_V \left\{ \frac{\partial}{\partial x}\left[\delta\Phi R_x \frac{\partial \Phi}{\partial x}\right] + \frac{\partial}{\partial y}\left[\delta\Phi R_y \frac{\partial \Phi}{\partial y}\right] - \left(\frac{\partial \delta\Phi}{\partial x}R_x\frac{\partial \Phi}{\partial x} + \frac{\partial \delta\Phi}{\partial y}R_y\frac{\partial \Phi}{\partial y}\right) \right.$$

$$\left. + \delta\Phi\left(B_x\frac{\partial \Phi}{\partial x} + B_y\frac{\partial \Phi}{\partial y} + \right) + \delta\Phi G\Phi + + \delta\Phi H \right\} dV = 0 \tag{9.6}$$

Green's theorem applied to the first two terms in Eq. 9.6 gives

$$\int_S \delta\Phi\left[R_x\frac{\partial \Phi}{\partial x}n_x + R_y\frac{\partial \Phi}{\partial y}n_y\right]dS - \int_V \left[\frac{\partial \delta\Phi}{\partial x}R_x\frac{\partial \Phi}{\partial x} + \frac{\partial \delta\Phi}{\partial y}R_y\frac{\partial \Phi}{\partial y}\right] dV$$

$$+ \int_V \left[\delta\Phi\left(B_x\frac{\partial \Phi}{\partial x} + B_y\frac{\partial \Phi}{\partial y} + \right) + \delta\Phi\left(G\Phi + H\right)\right] dV = 0.0 \tag{9.7}$$

To evaluate the surface integral, either we must know

$$q = R_x\frac{\partial \Phi}{\partial x}n_x + R_y\frac{\partial \Phi}{\partial y}n_y \tag{9.8}$$

or we must specify

$$\delta\Phi = 0 \tag{9.9}$$

The latter implies that we know the value of Φ. These two conditions correspond to the boundary conditions listed for the strong form. The final governing equation in its weak form is usually written as

$$\int_V \left(\frac{\partial \delta\Phi}{\partial x}R_x\frac{\partial \Phi}{\partial x} + \frac{\partial \delta\Phi}{\partial y}R_y\frac{\partial \Phi}{\partial y}\right) dV - \int_V \delta\Phi\left(B_x\frac{\partial \Phi}{\partial x} + B_y\frac{\partial \Phi}{\partial y}\right) dV - \int_V \delta\Phi G\Phi \, dV$$

$$= \int_V \delta\Phi H \, dV + \int_S \delta\Phi q \, dS \tag{9.10}$$

9.1.2 Second Weak Form. The second weak form is obtained when the volume integral associated with the parameters B_x and B_y is integrated by parts. Thus, from Eq. 9.7, we have

$$\int_V \left[\delta\Phi \left(B_x \frac{\partial \Phi}{\partial x} + B_y \frac{\partial \Phi}{\partial y} \right) \right] dV = \int_V \left[\frac{\partial}{\partial x} \left(\delta\Phi B_x \Phi \right) + \frac{\partial}{\partial y} \left(\delta\Phi B_y \Phi \right) - \frac{\partial \delta\Phi}{\partial x} B_x \Phi - \frac{\partial \delta\Phi}{\partial x} B_y \Phi \right] dV \quad (9.11)$$

Application of Green's theorem to the first two terms on the right side gives

$$\int_V \left[\delta\Phi \left(B_x \frac{\partial \Phi}{\partial x} + B_y \frac{\partial \Phi}{\partial y} \right) \right] dV = \int_S \delta\Phi \left[B_x \Phi n_x + B_y \Phi n_y \right] dS + \int_V \left[-\frac{\partial \delta\Phi}{\partial x} B_x \Phi - \frac{\partial \delta\Phi}{\partial x} B_y \Phi \right] dV \quad (9.12)$$

After substitution of Eq. 9.12 into Eq. 9.7, we obtain

$$\int_S \delta\Phi \left[\left(R_x \frac{\partial \Phi}{\partial x} + B_x \Phi \right) n_x + \left(R_y \frac{\partial \Phi}{\partial y} + B_y \Phi \right) n_y \right] dS - \int_V \left[\frac{\partial \delta\Phi}{\partial x} R_x \frac{\partial \Phi}{\partial x} + \frac{\partial \delta\Phi}{\partial y} R_y \frac{\partial \Phi}{\partial y} \right] dV$$

$$- \int_V \left[\frac{\partial \delta\Phi}{\partial x} B_x \Phi + \frac{\partial \delta\Phi}{\partial x} B_y \Phi \right] dV + \int_V \left[\delta\Phi \left(B\Phi + H \right) \right] dV = 0.0 \quad (9.13)$$

This version of the weak form requires different boundary conditions compared with the first version. Here, we must specify either

$$\tilde{q} = \left(R_x \frac{\partial \Phi}{\partial x} + B_x \Phi \right) n_x + \left(R_y \frac{\partial \Phi}{\partial y} + B_y \Phi \right) n_y \quad (9.14)$$

or

$$\delta\Phi = 0 \quad (9.15)$$

Both weak forms create nonsymmetric stiffness matrices for $B_x \neq 0$ and/or $B_y \neq 0$. The primary advantage of one form over the other is dependent on the user's knowledge of the boundary condition, that is, whether q is known or $\tilde{q}$ is known. For many engineering applications, q has more physical meaning and is more likely to be the boundary condition specified.

9.1.3 Axisymmetric Formulations. Up to this point, we have thought of our equation as describing a phenomenon taking place within a plane area, most likely with unit depth. Another type of two-dimensional analysis arises for problems that are axisymmetric. In these cases, dV takes on a different meaning as shown in Fig. 9.1.

The corresponding equation using the cylindrical coordinates r and z is

$$\frac{1}{r} \frac{\partial}{\partial r} \left[r R_r \frac{\partial \Phi}{\partial r} \right] + \frac{\partial}{\partial z} \left[R_z \frac{\partial \Phi}{\partial z} \right] + B_r \frac{\partial \Phi}{\partial r} + B_z \frac{\partial \Phi}{\partial z} + G\Phi + H = 0.0 \quad (9.16)$$

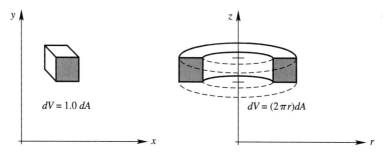

$dV = 1.0\, dA$

$dV = (2\pi r)\, dA$

Figure 9.1. Differential dV's.

Again, you should understand that there are other forms for this equation. We have once more assumed that placing the coefficients B_x and B_y outside the derivative creates the mathematical model of the physical phenomenon most users have in mind.

The weak form of the above equation is obtained as follows:

$$\int_V \delta\Phi \left\{ \frac{1}{r}\frac{\partial}{\partial r}\left[rR_r\frac{\partial\Phi}{\partial r}\right] + \frac{\partial}{\partial z}\left[R_z\frac{\partial\Phi}{\partial z}\right] + B_r\frac{\partial\Phi}{\partial r} + B_z\frac{\partial\Phi}{\partial z} + G\Phi + H \right\}\, dV = 0.0 \tag{9.17}$$

Integration by parts gives

$$\int_V \left\{ \frac{1}{r}\frac{\partial}{\partial r}\left[\delta\Phi\, rR_r\frac{\partial\Phi}{\partial r}\right] + \frac{\partial}{\partial z}\left[R_z\frac{\partial\Phi}{\partial z}\right] \right\}\, dV - \int_V \left\{ \frac{1}{r}\frac{\partial\delta\Phi}{\partial r} rR_r\frac{\partial\Phi}{\partial r} + \frac{\partial\delta\Phi}{\partial z} R_z\frac{\partial\Phi}{\partial z} \right\}\, dV$$

$$+ \int_V \delta\Phi \left\{ B_r\frac{\partial\Phi}{\partial r} + B_z\frac{\partial\Phi}{\partial z} + G\Phi + H \right\}\, dV = 0.0 \tag{9.18}$$

Use of the divergence theorem now gives

$$\int_S \left\{ \delta\Phi R_r\frac{\partial\Phi}{\partial r} n_r + \delta\Phi R_z\frac{\partial\Phi}{\partial z} n_z \right\}\, dS - \int_V \left\{ \frac{1}{r}\frac{\partial\delta\Phi}{\partial r} rR_r\frac{\partial\Phi}{\partial r} + \frac{\partial\delta\Phi}{\partial z} R_z\frac{\partial\Phi}{\partial z} \right\}\, dV$$

$$+ \int_V \left\{ \delta\Phi B_r\frac{\partial\Phi}{\partial r} + \delta\Phi B_z\frac{\partial\Phi}{\partial z} + \delta\Phi G\Phi + \delta\Phi H \right\}\, dV = 0.0 \tag{9.19}$$

On the surface,

$$R_r\frac{\partial\Phi}{\partial r} n_r + R_z\frac{\partial\Phi}{\partial z} n_z = q \tag{9.20}$$

Hence,

$$\int_S \delta\Phi q\, dS - \int_V \left\{ \delta\frac{\partial\Phi}{\partial r} R_r\frac{\partial\Phi}{\partial r} + \delta\frac{\partial\Phi}{\partial z} R_z\frac{\partial\Phi}{\partial z} \right\}\, dV$$

$$+ \int_V \left\{ \delta\Phi B_r\frac{\partial\Phi}{\partial r} + \delta\Phi B_z\frac{\partial\Phi}{\partial z} + \delta\Phi G\Phi + \delta\Phi H \right\}\, dV = 0.0 \tag{9.21}$$

This equation is identical in form to that used for the plane formulation. The difference lies in the specification for dV and dS, where now

$$dV = dr\,dz\,r\,d\Theta$$

$$dS = dS\,r\,d\Theta \tag{9.22}$$

If we take $d\Theta = 2\pi$ radians, then

$$dV = 2\pi r\,dr\,dz$$

$$dS = 2\pi r\,dS \tag{9.23}$$

(In order not to change our use of S to represent both a segment of surface and a segment along the line representing that surface, we have used dS on both the right-hand side and the left-hand side of the surface differentials. Note, however, that the left-hand dS represents a surface segment with dimensions of area, and the right-hand dS represents a segment along the line representing that surface with dimensions of length.) Therefore, to switch from a finite element program for the analysis of a two-dimensional problem in the x-y plane to one that is an axisymmetric problem in the r-z plane, we need only include the term $2\pi r$ as a multiplier of all integrands.

9.1.4 Mixed Boundary Conditions.
The term *mixed boundary condition* refers to situations where neither Φ nor q (or $\bar{q}$), but rather a linear relationship between the two, is known. An example is convective heat transfer, where the transfer of thermal energy across a boundary is dependent on the difference between the ambient temperature outside the boundary and the (yet to be determined) temperature at the surface of the region being analyzed. This can be expressed as

$$R_x\frac{\partial\Phi}{\partial x}n_x + R_y\frac{\partial\Phi}{\partial y}n_y = C_h(\Phi_a - \Phi) \tag{9.24}$$

Substitution of this expression into the surface integral of Eq. 9.7 gives

$$\int_S \delta\Phi\left[C_h(\Phi_a - \Phi)\right]dS = \int_S \delta\Phi C_h\Phi_a\,dS - \int_S \delta\Phi C_h\Phi\,dS \tag{9.25}$$

The integral containing the known ambient temperature is added to the right-hand side of the finite element equations, whereas the integral containing the yet-to-be-determined temperature on S is added to the stiffness matrix.

9.1.5 Necessary Conditions for a Unique Solution.
To obtain a unique solution to our problem it is necessary to specify, at every point of the surface, one of the boundary conditions mentioned above. Furthermore, if $G(x) = 0$, then at least one point on the surface must be used to specify a value for Φ. Such a specification, as discussed in Chapter 4, is sufficient to make the matrix nonsingular. A mixed boundary condition specification at one point is also sufficient to create a nonsingular matrix. It is left as a mental exercise for the reader to determine the whys and wherefores of this last statement.

9.2 THE FINITE ELEMENT APPROXIMATION

9.2.1 Review of Finite Element Equations and Notation. We are now ready to substitute into the weak form of our governing equation (see Eq. 9.10) the finite element approximation for Φ developed in the previous chapter. The related equations and notations developed are now reviewed.

Equations Related to the Volume Integrals

$$x = \lfloor N(u, v) \rfloor \{X\} \tag{9.26}$$

$$y = \lfloor N(u, v) \rfloor \{Y\} \tag{9.27}$$

$$\Phi(u, v) = \lfloor N(u, v) \rfloor \{\Phi\} \tag{9.28}$$

$$\begin{bmatrix} \partial x/\partial u & \partial x/\partial v \\ \partial y/\partial u & \partial y/\partial v \end{bmatrix} = \begin{bmatrix} \lfloor \partial N/\partial u \rfloor \{X\} & \lfloor \partial N/\partial v \rfloor \{X\} \\ \lfloor \partial N/\partial u \rfloor \{Y\} & \lfloor \partial N/\partial v \rfloor \{Y\} \end{bmatrix} \tag{9.29}$$

$$\left[J\left(\frac{u, v}{x, y} \right) \right] = \begin{bmatrix} \partial u/\partial x & \partial u/\partial y \\ \partial v/\partial x & \partial v/\partial y \end{bmatrix} \tag{9.30}$$

$$\left[J\left(\frac{x, y}{u, v} \right) \right] = \begin{bmatrix} \partial x/\partial u & \partial x/\partial v \\ \partial y/\partial u & \partial y/\partial v \end{bmatrix} \tag{9.31}$$

$$\left[J\left(\frac{u, v}{x, y} \right) \right] = \left[J\left(\frac{x, y}{u, v} \right) \right]^{-1} \tag{9.32}$$

$$\begin{bmatrix} \lfloor \partial N/\partial x \rfloor \\ \lfloor \partial N/\partial y \rfloor \end{bmatrix} = \left[J\left(\frac{u, v}{x, y} \right) \right]^{T} \begin{bmatrix} \lfloor \partial N/\partial u \rfloor \\ \lfloor \partial N/\partial v \rfloor \end{bmatrix} \tag{9.33}$$

$$\|J\| = \left\| \left[J\left(\frac{u, v}{x, y} \right) \right] \right\| \tag{9.34}$$

$$dA_{xy} = \|J\| \, dA_{uv} \tag{9.35}$$

Equations Related to the Surface Integrals

$$x = \lfloor N(u) \rfloor \{X\} \tag{9.36}$$

$$y = \lfloor N(u) \rfloor \{Y\} \tag{9.37}$$

$$\Phi(s) = \lfloor N(u) \rfloor \{\Phi\} \tag{9.38}$$

$$\begin{Bmatrix} dx/du \\ dy/du \end{Bmatrix} = \begin{Bmatrix} \lfloor dN/du \rfloor \{Y\} \\ \lfloor dN/du \rfloor \{X\} \end{Bmatrix} \tag{9.39}$$

$$\|J_s\| = \left[(dx/du)^2 + (dy/du)^2 \right]^{1/2} \tag{9.40}$$

$$dS = \|J_s\| \, du \tag{9.41}$$

9.2.2 The Volume Integrals. With the preceding approximations, the volume integrals appearing in the weak form of our governing equation (see Eq. 9.10) become

$$\int_{V_{xy}} \left[\frac{\partial \delta\Phi}{\partial x} R_x \frac{\partial \Phi}{\partial x} + \frac{\partial \delta\Phi}{\partial y} R_y \frac{\partial \Phi}{\partial y} \right] dV$$

$$= \int_{V_{xy}} \lfloor \partial\delta\Phi/\partial x \quad \partial\delta\Phi/\partial y \rfloor \begin{bmatrix} R_x & 0 \\ 0 & R_y \end{bmatrix} \begin{Bmatrix} \partial\Phi/\partial x \\ \partial\Phi/\partial y \end{Bmatrix} dV$$

$$= \lfloor \delta\Phi \rfloor \left[\int_{V_{xy}} \begin{bmatrix} \lfloor \partial N/\partial x \rfloor \\ \lfloor \partial N/\partial y \rfloor \end{bmatrix}^T \begin{bmatrix} R_x & 0 \\ 0 & R_x \end{bmatrix} \begin{bmatrix} \lfloor \partial N/\partial x \rfloor \\ \lfloor \partial N/\partial y \rfloor \end{bmatrix} dV \right] \{\Phi\}$$

$$= \lfloor \delta\Phi \rfloor \left[\int_{V_{uv}} \left[\begin{Bmatrix} \frac{\partial N}{\partial x} \end{Bmatrix} \quad \begin{Bmatrix} \frac{\partial N}{\partial y} \end{Bmatrix} \right] \begin{bmatrix} R_x & 0 \\ 0 & R_x \end{bmatrix} \begin{bmatrix} \lfloor \partial N/\partial x \rfloor \\ \lfloor \partial N/\partial y \rfloor \end{bmatrix} \|J\| \, dV \right] \{\Phi\}$$

$$= \lfloor \delta\Phi \rfloor [S_1] \{\Phi\} \tag{9.42}$$

$$\int_{V_{xy}} \left[\delta\Phi \left(B_x \frac{\partial\Phi}{\partial x} + B_y \frac{\partial\Phi}{\partial y} \right) \right] dV = \int_{V_{xy}} \delta\Phi \left\lfloor B_x \quad B_y \right\rfloor \begin{Bmatrix} \partial\Phi/\partial x \\ \partial\Phi/\partial y \end{Bmatrix} dV$$

$$= \lfloor \delta\Phi \rfloor \left[\int_{V_{xy}} \{N\} \lfloor B_x \quad B_y \rfloor \begin{bmatrix} \lfloor \partial N/\partial x \rfloor \\ \lfloor \partial N/\partial y \rfloor \end{bmatrix} dV \right] \{\Phi\} \tag{9.43}$$

$$= \lfloor \delta\Phi \rfloor \left[\int_{V_{uv}} \{N\} \lfloor B_x \quad B_y \rfloor \begin{bmatrix} \lfloor \partial N/\partial x \rfloor \\ \lfloor \partial N/\partial y \rfloor \end{bmatrix} \|J\| dV \right] \{\Phi\}$$

$$= \lfloor \delta\Phi \rfloor [S_2] \{\Phi\}$$

$$\int_{V_{xy}} \delta\Phi G\Phi \, dV = \lfloor \delta\Phi \rfloor \left[\int_{V_{xy}} \{N\} G \lfloor N \rfloor dV \right] \{\Phi\}$$

$$= \lfloor \delta\Phi \rfloor \left[\int_{V_{uv}} \{N\} G \lfloor N \rfloor \|J\| dV \right] \{\Phi\} \tag{9.44}$$

$$= \lfloor \delta\Phi \rfloor [S_3] \{\Phi\}$$

$$\int_{V_{xy}} \delta\Phi H \, dV = \lfloor \delta\Phi \rfloor \left[\int_{V_{xy}} \{N\} H \, dV \right]$$

$$= \lfloor \delta\Phi \rfloor \left[\int_{V_{uv}} \{N\} H \|J\| dV \right] \{\Phi\} \tag{9.45}$$

$$= \lfloor \delta\Phi \rfloor \{f_H\}$$

9.2.3 The Surface Integrals. When the finite element approximations are substituted into the surface integrals that appear in the weak form of our governing equation, they become

$$\int_S \delta\Phi q \, dS = \lfloor\delta\Phi\rfloor \int_S \{N\}q \, dS$$

$$= \lfloor\delta\Phi\rfloor \int_u \{N\}q\|J_s\| \, du \tag{9.46}$$

$$= \lfloor\delta\Phi\rfloor\{f_q\}$$

$$\int_S \delta\Phi C_h\Phi_a \, dS = \lfloor\delta\Phi\rfloor \int_S \{N\}C_h\Phi_a \, dS$$

$$= \lfloor\delta\Phi\rfloor \int_u \{N\}C_h\Phi_a\|J_s\| \, du \tag{9.47}$$

$$= \lfloor\delta\Phi\rfloor\{f_a\}$$

$$\int_S \delta\Phi C_h\Phi \, dS = \lfloor\delta\Phi\rfloor \int_S \{N\}C_h\lfloor N\rfloor\{\Phi\} \, dS$$

$$= \lfloor\delta\Phi\rfloor \int_u \{N\}\, C_h \lfloor N\rfloor\|J_s\| \, du \,\{\Phi\} \tag{9.48}$$

$$= \lfloor\delta\Phi\rfloor[S_h]\{\Phi\}$$

9.2.4 Final FEM Equation. The final finite element equation is

$$[K]\{\Phi\} = \{Q\} \tag{9.49}$$

where

$$[K] = [S_1] - [S_2] - [S_3] + [S_h] \tag{9.50}$$

$$\{Q\} = \{f_H\} + \{f_q\} + \{f_h\} \tag{9.51}$$

and

$$[S_1] = \int_{V_{uv}} \left[\left\{\frac{\partial N}{\partial x}\right\} \quad \left\{\frac{\partial N}{\partial y}\right\} \right] \begin{bmatrix} R_x & 0 \\ 0 & R_x \end{bmatrix} \begin{bmatrix} \lfloor\partial N/\partial x\rfloor \\ \lfloor\partial N/\partial y\rfloor \end{bmatrix} \|J\| \, dV \tag{9.52}$$

$$[S_2] = \int_{V_{uv}} \{N\} \lfloor B_x \quad B_y \rfloor \begin{bmatrix} \lfloor \partial N / \partial x \rfloor \\ \lfloor \partial N / \partial y \rfloor \end{bmatrix} \|J\| \, dV \tag{9.53}$$

$$[S_3] = \int_{V_{uv}} \{N\} G \lfloor N \rfloor \|J\| \, dV \tag{9.54}$$

$$[S_h] = \int_u \{N\} C_h \lfloor N \rfloor \|J_s\| \, du \tag{9.55}$$

$$\{f_H\} = \int_{V_{uv}} \{N\} H \|J\| \, dV \tag{9.56}$$

$$\{f_q\} = \int_u \{N\} q \|J_s\| \, du \tag{9.57}$$

$$\{f_a\} = \int_u \{N\} C_h \Phi_a \|J_s\| \, du \tag{9.58}$$

9.3 GAUSSIAN QUADRATURE

The volume and surface integrals in the weak form of our problem must be integrated element by element in our finite element code. As indicated above, the integration takes place over the parent element in the u-v plane. The integration is performed using Gaussian quadrature as was done for our second-order ordinary differential equations in Chapter 4. For all cases, we use the general form

$$\int_{V_{uv}} f(u, v) \, dV = \sum_{i=1}^{n} W_i f(u_i, v_i) \tag{9.59}$$

where n equals the number of quadrature points. For the case of square elements in the u-v plane, the points and weights can be obtained using the one-dimensional weights and points, that is,

$$\int_{V_{uv}} f(u, v) \, dV = \int_{v=-1}^{+1} \left[\int_{u=-1}^{+1} f(u, v) \, du \right] dv \tag{9.60}$$

$$\approx \sum_{i=1}^{n} \sum_{j=1}^{n} W_i W_j f(u_i, v_j)$$

where n now equals the number of points in one direction; hence, the total number of points is n^2.

Following is a table of Gaussian points and weights for square elements in the u-v plane. Note that these points are for $-1 \le u \le +1$ and $-1 \le v \le +1$, which is the same domain that was used in the previous chapter to define the shape functions for quadrilateral elements.

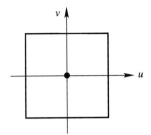

u	v	W
0	0	4

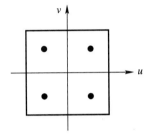

u	v	W
$-a$	$-a$	1
a	$-a$	1
a	a	1
$-a$	a	1

$a = 1/\sqrt{3}$

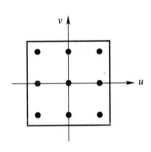

u	v	W
$-a$	$-a$	w_1
0	$-a$	w_2
a	$-a$	w_1
$-a$	0	w_2
0	0	w_3
a	0	w_2
$-a$	a	w_1
0	a	w_2
a	a	$w1$

$a = \sqrt{0.6}$
$w_1 = 25/81$
$w_2 = 40/81$
$w_3 = 64/81$

For triangular elements, the point and weights we use are

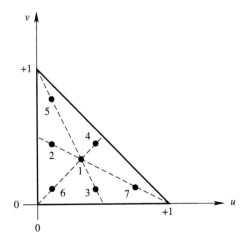

Quadrature Point	u	v	z	W
1	0.333333	0.333333	0.333333	0.112500
2	0.059716	0.470142	0.470142	0.066197
3	0.470142	0.059716	0.470142	0.066197
4	0.470142	0.470142	0.059716	0.066197
5	0.101287	0.797427	0.101287	0.062970
6	0.101287	0.101287	0.797427	0.062970
7	0.797427	0.101287	0.101287	0.062970

Here a third, dependent, coordinate has been added such that

$$u + v + z = 1.0 \tag{9.61}$$

These three coordinates coincide with area coordinates for triangles. It is these coordinates that are shown in most of the published tables for the quadrature in triangular regions. However, the u and v coordinates can be interpreted as either Cartesian coordinates or area coordinates. Refer to Appendix C for a more complete discussion of this, as well as a more complete discussion of integration in two dimensions.

For the integration of the surface integrals, the integrands are evaluated on the sides of the parent element. Hence, the quadrature points must likewise be located along the sides. This corresponds to the one-dimensional quadrature used in ode2.m, and those quadrature coordinates and weights can be used for this purpose.

9.4 PROGRAMMING PRELIMINARIES

The programming details for our new code are very similar to those given for poisson.m in Chapter 6. It might be worthwhile to return to that section for review.

9.4.1 Integration of $[K]_e$ and $\{f_H\}_e$ Matrices. The integration of the stiffness matrix and the right-hand-side array is done by Gaussian quadrature. Thus, each integrand must be evaluated at each quadrature point in each element or along each surface segment. As part of these evaluations, each shape function and the derivative of each shape function with respect to the u and v coordinates in the parent element, and each shape function on a side and its derivative with respect to its S coordinate, must be evaluated. That is, it is necessary to have for each element and each side, at each quadrature point, for each shape function, the value of

$$N, \quad \frac{\partial N}{\partial u}, \quad \frac{\partial N}{\partial v}, \quad N_s, \quad \frac{\partial N_s}{\partial u}$$

However, these values are associated only with the parent element and its sides; hence, they are identical for each element. Therefore, the values are calculated only once and stored in a three-dimensional array. In addition, the required number of quadrature points and their corresponding weights must be available for the integrations. All of this information is obtained at the beginning of the code from the function

```
[SF, WT, NUMQPT, NPSIDE] = SFquad{NNPE}
```

where

SF(I,J,K)	Shape function array (see below for explanation)
WT(1,I)	Weight for area (volume) integration, quadrature point I
WT(2,I)	Weight for line (surface) integration, quadrature point I
NUMQPT(1)	Number of quadrature points for area integration
NUMQPT(2)	Number of quadrature points for line integration
NPSIDE(I,J)	Element node number for Jth node on side I (see below)
NNPE	Number of nodes per element

It is necessary to supply SFquad only with the type of element (NNPE), and it will supply the information necessary to perform the Gaussian quadrature. For NNPE = 6, the following arrays will be returned to the main program:

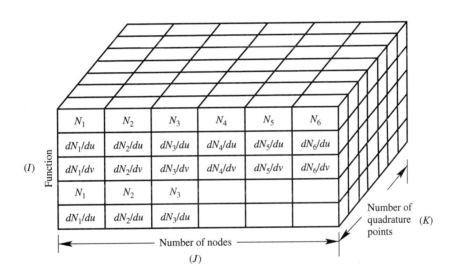

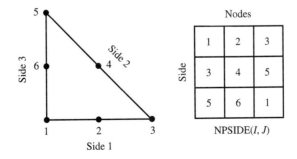

NUMQPT(I)	WT(I,J)						
7	0.112	0.066	0.066	0.066	0.063	0.063	0.063
3	0.556	0.889	0.556	.	.	.	.

In the SF array, the first three layers (I = 1, 2, 3) refer to the two-dimensional shape functions, and the last two layers (I = 4, 5) refer to the one-dimensional shape functions. For example,

SF(2,3,4) contains the derivative with respect to u, of the two-dimensional shape function associated with node 3, evaluated at quadrature point 4, i.e.,

$$\frac{\partial N_3(u_4, v_4)}{\partial u}$$

SF(4,2,3) contains the one-dimensional shape function associated with node 2, evaluated at quadrature point 3, i.e.,

$$N_2(u_3)$$

An explanation and complete listing of **SFquad.m** are given in Appendix B. There you will find how it is possible to change the number of quadrature points used for a given element and even to add new elements such as the nine-node Lagrangian element. The reader is encouraged to study this material and to experiment by making such changes.

9.4.2 The Global Stiffness Matrix. The stiffness matrix will be nonsymmetric for problems where G is not identically zero. This means that the matrix must be stored in the manner described in Chapter 3 and Appendix A for nonsymmetric, banded matrices.

9.4.3 Boundary Specifications. The specification of the boundary conditions is similar to that used in our previous codes; that is, NPBC = 0 means PHI is unkown but the flux is known and specified, whereas NPBC = 1 means PHI is known and specified but the flux is unknown. If PHI is unknown, the flux can be specified in the following three ways:

1. Q(I) = nodal value of a point flux or source
2. QS(I) = nodal value of a distributed flux or source, q_s
3. HS(I) = nodal value of the coefficient for a convective type flux, $q_x = C_h(\Phi_a - \Phi)$

These values are initialized by the code to have a default value of zero. The user can then change their values in INITIAL.m. As with the other codes, the nodal values of **NPcode** can be used in INITIAL.m to determine the correct boundary conditions to be assigned to any node. When a convective-type flux is specified by a nonzero HS value, a corresponding ambient PHI value must also be specified. This is done by placing the ambient value in the PHI array for these nodes. The code can use these values as the ambient values because, under these boundary specifications, NPBC will be equal to zero and PHI will not represent a known boundary temperature. For nonzero values of either QS or HS, a surface quadrature will be made along the sides of the elements associated with these nodes. The quadrature values for QS and HS and ambient PHI will be obtained by using a finite element interpolation of the quantities based on their nodal values.

9.4.4 User-Provided INCLUDE Codes. The user INCLUDE codes are the same for our new code as they were for poisson.m. Now, however, in COEF.m it is necessary to define the following parameters: R_x, R_y, B_x, B_y, G, and H.

9.4.5 DATA Files. The data files necessary to run the new code are identical to those required for poisson.m, and the user is advised to consult Chapter 7 for review.

9.4.6 Axisymmetric Analyses. The code will ask the user to define the type of analysis to be conducted:
   2-D rectangular analysis
   Axisymmetric about the x axis
   Axisymmetric about the y axis

Note: For the axisymmetric analyses, there cannot be any nodal points on the negative side of the axis of symmetry.

9.5 PROGRAM steady.m

Program steady.m is the finite element code for solving Eq. 9.1. In addition to steady.m, you will need the following files in your working directory:

Input Data	User's INCLUDEs	Supplied Functions
MESHo	INITIAL.m	SFquad.m
NODES	COEF.m	nGAUSS.m
NP		
NWLD		

The notation used in steady.m for the global FEM equation is

$$[SK]\{PHI\} = \{RHS\}$$

9.5.1 Flow Chart

LOAD files from mesh.m and newnum.m

User-written **INITIAL.m**
to initialize problem parameters

Function: **SFquad.m** to create SF array

FOR all elements in mesh

FOR all quadrature points in element

User-written **COEF.m**

Determine quadrature contribution to [S] and {QE}

END loop over quadrature points

FOR all sides of element

FOR all quadrature points on surface

Calculate contributions to [S] and {QE}

END loop over quadrature points
END loop over sides
END loop over elements

Assemble [S] and {QE} into global matrices

Specify known values of PHI

Function: **nGAUSS.m** to solve [SK]{PHI} = {RHS}

Save {PHI} values

9.5.2 Code

```
%*********************
%   PROGRAM steady.m
%*********************

    clear
%-----------------------
%  LOAD DATA
%-----------------------
    load MESHo    -ASCII
    load NODES    -ASCII
    load NP       -ASCII
    load NWLD     -ASCII

    NUMNP = MESHo(1);
    NUMEL = MESHo(2);
    NNPE  = MESHo(3);

    for I=1:NUMNP;
      XORD(I)=NODES(I,1);
      YORD(I)=NODES(I,2);
      NPcode(I)=NODES(I,3);
    end

%  --------------------
%  Determine bandwidth and
%  diagonal column for
%  nonsymmetric storage
%  --------------------
    IB    = NWLD(NUMNP+1);
    IDIAG=IB;
    IB    = 2*IB-1;


%  --------------------
%  General Initialization
%  --------------------
    PI=4.0*atan(1.0);
    if NNPE == 3
       NSPE=3;
       NNPS=2;
    elseif NNPE == 6
       NSPE=3;
       NNPS=3;
    elseif NNPE == 4
       NSPE=4;
       NNPS=2;
    elseif NNPE == 8
       NSPE=4;
       NNPS=3;
    end
```

Load data from mesh.m and newnmu.m.
All four files must be in working directory.

File	Contents
MESHo	NUMNP, NUMEL, NNPE
NODES	XORD, YORD, NPBC
NP	NP array
NWLD	New nodal numbers
	Last entry is IB

Variable	Definition
NUMNP	Number of nodal points
NUMEL	Number of elements
NNPE	Number of nodes per element
XORD	Node's x coordinates
YORD	Node's y coordinates
NPcode	User's identification code for each node

Convert IB from newnum.m to IB for a
nonsymmetric matrix. IDIAG is column
number for diagonal terms. See Appendix B
for addition information.

In MATLAB, pi $= \pi$, but in other languages
it must be set.

Variable	Definition
NNPE	Number of nodes per element
NSPE	Number of sides per element
NNPS	Number of nodes per side

```
    for I=1:NUMNP;
        Q(I)=0;
        QS(I)=0;
        HS(I)=0;
        PHI(I)=0;
        NPBC(i)=0.0;
    end

    for I=1:NUMNP
        for J=1:IB
            SK(I,J)=0;
        end
    end
```

```
% -----------------------------------
% Ask user for type of analysis desired
% -----------------------------------

disp(' ')
disp('ENTER:')
disp('-----------------------------')
disp('0 for 2D rectangular'          )
disp('1 for axi-symmetric about x-axis')
disp('2 for axi-symmetric about y-axis')
disp('-----------------------------')
IRZ = input(' < ');

    if IRZ < 0 | IRZ > 2
        disp('--------------------')
        disp('ERROR IN INPUT '        )
        disp('IRZ has been set to 0')
        disp('--------------------')
        IRZ = 0;
    end
```

```
% -----------------------------------
% Get shape function quadrature data
% -----------------------------------
    [SF,WT,NUMQPT,NPSIDE] = SFquad(NNPE);
```

```
% -----------------------------------
% Include user-written initialization
% -----------------------------------
    INITIAL
```

Variable	Definition
Q	Q of $[K]\{\Phi\} = \{Q\}$
QS	Nodal values of surface flux, q_s
HS	Nodal value of C_h in Eq. 9.24
PHI	Nodal values of Φ
NPBC	Nodal point boundary condition
SK	Global stiffness matrix

steady.m can analyze:
(1) Two-dimensional problems in rectangular coordinates.
(2) Problems that are axisymmetric about the x axis, hence two-dimensional.
(3) Problems that are axisymmetric about the y axis, hence two-dimensional.

If user does not enter 0, 1, or 2, default to 0.

SFquad.m is the function that initializes the shape factor array, SF, according to the element type specified by NNPE, the number of nodes per element. The values are calculated at each quadrature point to expedite the numerical integration. See Appendix D for complete details.

INITIAL.m: User-written INCLUDE code to set all boundary conditions using the NPcode numbers for each node or other user-written logic.

```
%---------------------------------
% Place Q in RHS, compact storage
%---------------------------------
   for I=1:NUMNP
     RHS(NWLD(I))=Q(I);
   end

% -----------------------
% Create element matrices
% -----------------------
   for I=1:NUMEL;
     LMENT=I;
     for J=1:NNPE;
       QE(J)=0.0;
       for K=1:NNPE;
          S(J,K)=0;
        end
      end

% -------------------------------
% Begin quadrature for each element
% -------------------------------
   JEND=NUMQPT(1);
   for J=1:JEND;
     XJ=0;
     YJ=0;
     RJAC(1,1)=0;
     RJAC(1,2)=0;
     RJAC(2,1)=0;
     RJAC(2,2)=0;
%    -------------------------------
%    Determine coordinate and Jacobian
%    -------------------------------
     for K=1:NNPE;
       NPK=NP(I,K);
       XJ=XJ+SF(1,K,J)*XORD(NPK);
       YJ=YJ+SF(1,K,J)*YORD(NPK);

       RJAC(1,1)=RJAC(1,1)+...
                 SF(2,K,J)*XORD(NPK);
       RJAC(1,2)=RJAC(1,2)+...
                 SF(3,K,J)*XORD(NPK);
       RJAC(2,1)=RJAC(2,1)+...
                 SF(2,K,J)*YORD(NPK);
       RJAC(2,2)=RJAC(2,2)+...
                 SF(3,K,J)*YORD(NPK);
     end

     DETJ=RJAC(1,1)*RJAC(2,2)...
         -RJAC(2,1)*RJAC(1,2);
```

Place specified nodal sources on right-hand side using new numbering for compact storage.

Begin compiling global stiffness matrix element by element.

Initialize element right-hand side.

Initialize element stiffness matrix.

Begin Gaussian quadrature for current element.

Initialize:
XJ and YJ: coordinates of quadrature point
RJAC(I,J): Jacobian matrix

Determine:

$$x = \lfloor N \rfloor \{X\}$$

$$y = \lfloor N \rfloor \{Y\}$$

$$\left[J \frac{(x, y)}{u, v)} \right] = \begin{bmatrix} \dfrac{\partial x}{\partial u} & \dfrac{\partial x}{\partial v} \\ \dfrac{\partial y}{\partial u} & \dfrac{\partial y}{\partial v} \end{bmatrix}$$

$$= \begin{bmatrix} \lfloor \partial N / \partial u \rfloor \{X\} & \lfloor \partial N / \partial v \rfloor \{X\} \\ \lfloor \partial N / \partial u \rfloor \{Y\} & \lfloor \partial N / \partial v \rfloor \{Y\} \end{bmatrix}$$

$$\text{DETJ} = \begin{Vmatrix} \dfrac{\partial x}{\partial u} & \dfrac{\partial x}{\partial v} \\ \dfrac{\partial y}{\partial u} & \dfrac{\partial y}{\partial v} \end{Vmatrix}$$

```
    if DETJ <=    0
      fprintf(1,'\n------------------ ')
      fprintf(1,'\n Error in steady.m ')
      fprintf(1,'\n DETJ =%7e',DETJ    )
      fprintf(1,'\n must be > 0.0'      )
      fprintf(1,'\n----------------\n')
      error
    end


%  ----------------------------
%  Determine inverse of Jacobian
%  ----------------------------
    RJACI(1,1)=+RJAC(2,2)/DETJ;
    RJACI(1,2)=-RJAC(1,2)/DETJ;
    RJACI(2,1)=-RJAC(2,1)/DETJ;
    RJACI(2,2)=+RJAC(1,1)/DETJ;



%  --------------------------------
%  Determine derivative of shape
%     functions in X-Y plane
%  --------------------------------
    for K=1:NNPE;
       DNDX(K)=RJACI(1,1)*SF(2,K,J)+...
               RJACI(2,1)*SF(3,K,J);
       DNDY(K)=RJACI(1,2)*SF(2,K,J)+...
               RJACI(2,2)*SF(3,K,J);
    end



%  ----------------------------------
%  Include user-written coefficients
%  RXJ, RYJ, BXJ, BYJ, GVJ, HVJ
%  ----------------------------------
    COEF



%  ------------------------------
%  Create element stiffness matrix
%  ------------------------------
    DV=DETJ;
    if IRZ == 1
       DV=2.0*PI*YJ*DV;
    elseif IRZ == 2
       DV=2.0*PI*XJ*DV;
    end
```

The sign of the determinant of the Jacobian matrix is based on counterclockwise numbering of the element nodes in the NP array. If this has been followed, and the determinant is less than or equal to zero, the mapping from parent element to current element is not proper. The current element has either been wrongly defined are is too distorted to use. Program is stopped.

$$\begin{bmatrix} \dfrac{\partial u}{\partial x} & \dfrac{\partial u}{\partial y} \\[2mm] \dfrac{\partial v}{\partial x} & \dfrac{\partial v}{\partial y} \end{bmatrix} = \begin{bmatrix} \dfrac{\partial x}{\partial u} & \dfrac{\partial x}{\partial v} \\[2mm] \dfrac{\partial y}{\partial u} & \dfrac{\partial y}{\partial v} \end{bmatrix}^{-1}$$

$$\begin{bmatrix} \dfrac{\partial N}{\partial x} \\[2mm] \dfrac{\partial N}{\partial y} \end{bmatrix} = \begin{bmatrix} \dfrac{\partial u}{\partial x} & \dfrac{\partial v}{\partial x} \\[2mm] \dfrac{\partial u}{\partial y} & \dfrac{\partial v}{\partial y} \end{bmatrix} \begin{bmatrix} \dfrac{\partial N}{\partial u} \\[2mm] \dfrac{\partial N}{\partial v} \end{bmatrix}$$

COEF.m is a user INCLUDE code that defines RX, RY, BX, BY, G, and H at the current quadrature point. See test problem for an example of COEF.m.

Account for type of analysis: (IRZ = 0) rectangular, (IRZ = 1) axisymmetric about x axis, and (IRZ = 2) axisymmetric about y axis.

```
    for K=1:NNPE;

      QE(K)=QE(K)+WT(1,J)*SF(1,K,J)*HVJ*DV;

      SFK=SF(1,K,J);
      for L=1:NNPE;
        SFL=SF(1,L,J);

          S(K,L)=S(K,L) +  WT(1,J)*(...
            DNDX(K)*RXJ*DNDX(L) ...
            +DNDY(K)*RYJ*DNDY(L) ...
            -SFK*BXJ*DNDX(L) ...
            -SFK*BYJ*DNDY(L) ...
            -SFK*GVJ*SFL)*DV;
      end
    end

  end
% ------------ end of volume quadrature


% ----------------------
% Begin surface quadrature
% ----------------------
  for J=1:NSPE
    CHKH=1;
    CHKQ=1;
    for K=1:NNPS;
      J1=NP(I,NPSIDE(J,K));
      CHKH=CHKH*HS(J1);
      CHKQ=CHKQ*QS(J1);
    end


    if CHKH ~= 0 |  CHKQ ~= 0
      KEND=NUMQPT(2);
      for K=1:KEND;
        YK   =0.0;
        XK   =0.0;
        dxdu =0.0;
        dydu =0.0;
        PHIK =0.0;
        HSK  =0.0;
        QSK  =0.0;
```

At this point in the code we are in element *I* at quadrature point *J*. All parameters and terms that appear in the integrand for the stiffness matrix and the right-hand side associated with the volume integration have been defined or calculated. These are

Integrand	Equation
$[N']^T[R][N']$	9.52
$\{N\}\lfloor B\rfloor[N']$	9.53
$\{N\}G\lfloor N\rfloor$	9.54
$\{N\}H$	9.56

We are still in element *I* and are ready to search all of its sides to determine if any one of them represents an exterior surface for which a numerical integration must be performed. The check is made by examining the nodal point values of q_s and h, at all nodes associated with each side. A single nodal value of zero for q_s, or h indicates that quantity should not be integrated.

Search each side. NSPE = number of sides per element.

Inspect each node. NNPS = number of nodes per side.

If either CHKH or CHKQ is nonzero, then a surface integration by Gaussian quadrature is begun. If not, then the next side is examined until all sides have been considered.

Initialize variables used in quadrature.

Variable	Definition
XK, YK	*x* and *y* coordinates
dxdu, dydu	*dx/du* and *dy/du*
PHIK	Φ_a in Eq. 9.24
HSK	C_h in Eq. 9.24
QSK	q^* in Eq. 9.2

```
    for L=1:NNPS;
      L1=NPSIDE(J,L);
      NPL=NP(I,L1);
      XK    =XK   +SF(4,L,K)*XORD(NPL);
      YK    =YK   +SF(4,L,K)*YORD(NPL);
      dxdu =dxdu +SF(5,L,K)*XORD(NPL);
      dydu =dydu +SF(5,L,K)*YORD(NPL);
      QSK   =QSK  +SF(4,L,K)*QS(NPL);
      HSK   =HSK  +SF(4,L,K)*HS(NPL);
      PHIK =PHIK +SF(4,L,K)*PHI(NPL);
    end
    dsdu = sqrt(dxdu^2 + dydu^2);
    duds = 1/dsdu;

    DS=dsdu;
    if IRZ == 1
        DS=2.0*PI*YK*DS;
    elseif IRZ == 2
        DS=2.0*PI*XK*DS;
    end

    for L=1:NNPS;
      L1=NPSIDE(J,L);
      if CHKQ ~= 0
        QE(L1)=QE(L1)+WT(2,K)*...
                    SF(4,L,K)*QSK*DS;
      end


      if CHKH ~= 0
        QE(L1)=QE(L1)+WT(2,K)*...
              SF(4,L,K)*HSK*PHIK*DS;
        for M=1:NNPS;
          M1=NPSIDE(J,M);
          S(L1,M1)=S(L1,M1) + WT(2,K) ...
              *SF(4,L,K)*HSK*SF(4,M,K)*DS;
        end
        end

      end

    end % surface quadrature

  end % if-statement for quadrature

end % loop over element sides
```

CHKQ and/or CHKH indicates that numerical integration of one or both of the following integrals is needed:

$$\int \{N_s\} q(s)\, ds \qquad \int \{N_s\} C_h(\Phi_a - \Phi)\, ds$$
(Eq. 9.57)   (Eqs. 9.58 and 9.55)

Determine quadrature values of

$$x \quad y \quad dx/du \quad dy/du \quad q_s \quad h \quad \Phi_a$$

$$ds/du = \sqrt{(dx/du)^2 + (dy/du)^2}$$

$$du/ds = (ds/du)^{-1}$$

Calculate differential surface area.

$$\int \{N_s\} q\, ds$$

$$\int \{N_s\} h\Phi_a\, ds$$

$$\int \{N_s\} h\lfloor N_s \rfloor\, ds$$

```
%-------------------------------
% Place completed element matrix
% in global SK and Q matrices
%-------------------------------
    for J=1:NNPE
      JNP=NP(I,J);
      JEQ=NWLD(JNP);
      RHS(JEQ)=RHS(JEQ)+QE(J);
      for K=1:NNPE
          KNP=NP(I,K);
          KEQ=NWLD(KNP);
          KB=(KEQ-JEQ)+IDIAG;
          SK(JEQ,KB)=SK(JEQ,KB)+S(J,K);
      end
    end

  end % Loop over elements


%  -------------------------
%  Specify known values of PHI
%  -------------------------
    for I=1:NUMNP;
      if NPBC(I) == 1
        NI=NWLD(I);
        SK(NI,IDIAG)=SK(NI,IDIAG)*1.0E+06;
        RHS(NI)=PHI(I)*SK(NI,IDIAG);
      end
    end

%  -----------------------------
%  Call Gauss elimination routine
%  -----------------------------
    PHI = nGAUSS(SK,RHS,NUMNP,IB);

%  -----------------------------
%  Renumber solution values
%  using original node numbers
%  -----------------------------
    for I=1:NUMNP;
        NI=NWLD(I);
        RHS(I)=PHI(NI);
    end
    for I=1:NUMNP
        PHI(I)=RHS(I);
    end

    save PHI PHI -ASCII
```

JNP = global equation number for row *J* of element stiffness matrix

JEQ = new row number used for bandwidth reduction (see Appendixes A and C)

KNP = global equation number for column *K* of element stiffness matrix

KEQ = new column number used for bandwidth reduction

KB = column number in banded storage (see Appendix A)

Decouple equations representing known Φ values by "blasting the diagonal."

Call equation solver, nGAUSS.m, for banded nonsymmetric matrices (see Appendix A).

Put Φ values in order of original numbering.

Save Φ values in file PHI.

end of **steady.m**

9.5.3 Test Problem. For our test problem, we again use the fact that whenever our finite element approximation can duplicate an exact solution, it will do so. For program **steady.m** we want a test problem where all the coefficients are nonzero and at least some vary from point to point. Thus, we have selected

$$R_x = 2xy \qquad R_y = 3xy$$

$$B_x = 2 \qquad B_y = 2$$

$$G = -4 \qquad H = -17x + 16y + 78$$

With these coefficients, the exact solution to our governing equation is

$$\Phi = 20 - 2x + 3y$$

To continue our test, we wish to define a rather irregular domain and use of one our new higher-order elements. Hence, we select the four-node element and the domain shown in the following figure:

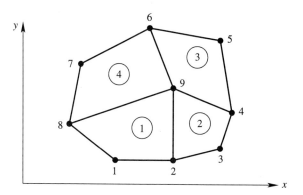

The data files corresponding to this mesh and the **NWLD** file are as follows:

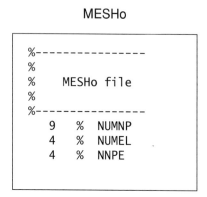

	MESHo	
%--------------		
%		
%	MESHo file	
%		
%--------------		
9	%	NUMNP
4	%	NUMEL
4	%	NNPE

	NP		
%--------------------			
%	NP array		
%			
% NP1	NP2	NP3	NP4
%--------------------			
1	2	9	8
2	3	4	9
4	5	6	9
9	6	7	8

NODES   NWLD

```
%------------------------        %------------------
%        NODES                   %   NWLD array
%                                %------------------
%  XORD  YORD  NPcode                   3
%------------------                     4
        8     2     1                   5
       13     2     1                   1
       17     3     1                   2
       18     6     1                   6
       17    12     1                   7
       11    13     1                   8
        5    10     1                   9
        4     5     1                   9   % IB
       13     8     0
```

Note that all nodes except the center node are given NPcode values of 1. In INITIAL.m these nodes will be specified as having known values of Φ. This is the easiest specification to make, and if the solution for the one remaining node is correct, then the code has passed a fairly good test. Note, however, that because of this specification, the surface integration used in the code has not been tested.

The new numbers listed in the NWLD file were selected randomly. The bandwidth given is the maximum possible for any new numbering.

The user INCLUDE codes for the test problem are as follows:

Initial.m

```
%-------------------------
% INITIAL.m
%
%  Set problem parameters.
%-------------------------

%-----------------------------------
%  This is written for the test
%  problem for which the exact
%  solution is:
%        PHI = 20 - 2x + 3y
%  Only node 9 is left unspecified.
%-----------------------------------

   for Ix=1:NUMNP
    if NPcode(Ix) == 1
      NPBC(Ix)=1;
      PHI(Ix) = 20-2*XORD(Ix)+3*YORD(Ix);
    else
      NPBC(Ix)=0;
    end
   end

%------------------------------------
% Print out exact answer for node 9
%------------------------------------
   exact = 20 - 2*XORD(9) + 3*YORD(9)
%------------------------------------
```

This code searches through the NPcode array to find those nodes on the boundary. Once these are found, NPBC is set equal to 1, and the correct value for Φ is calculated using the node's coordinates. Before leaving INITIAL.m, the value of Φ at node 9 (the unspecified node) is calculated and printed to the screen so that the user can verify the answer given by steady.m.

COEF.m

```
%----------------------------------------------------------*
%                                                          *
%    d    d@    d    d@       d@        d@                  *
%   --(RX--) + --(RY--) + BX-- + BY-- + G@ + HV = 0         *
%   dx   dx    dy   dy       dx        dy                   *
%                                                          *
%----------------------------------------------------------*


%-------------------------------------
%   Coefficients for solution to be:
%           PHI = 20 - 2x + 3y
%-------------------------------------

     RXJ = 2*XJ*YJ;
     RYJ = 3*XJ*YJ;

     BXJ=2;
     BYJ=2;

     GVJ = -4;
     HVJ = -17*XJ + 16*YJ + 78;
```

COEF.m uses the coordinates of the current quadrature point, XJ and YJ, to calculate the specified values for all parameters.

The solution will be saved in your working directory in file PHI. The values will be

$$10.0 \quad 0.0 \quad -5.0 \quad 2.0 \quad 22.0 \quad 37.0 \quad 40.0 \quad 27.0 \quad 18.0$$

As with poisson.m, three auxiliary codes are helpful in using steady.m. These can be found in Appendix D and are

mesh.m	Generates mesh files MESHo, NODES, NP
newnum.m	Generates data file NWLD
topo.m	Creates plots of PHI values

topo.m can be used to obtain the following contour plot of the solution:

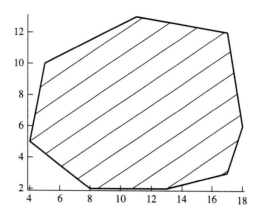

EXERCISES

Study Problems

S1. For an application (e.g., heat transfer) of your choice, state the physical significance of the boundary conditions as specified by the two different weak forms given in this chapter.

S2. For the differential equation

$$\frac{d^2 y}{dx^2} = 12.0$$

with boundary conditions

$$3.0(y - 2.0) = \frac{dy}{dx} \quad \text{at } x = 0$$

and

$$\frac{dy}{dx} = 15.0 \quad \text{at } x = 2$$

is there sufficient information to determine a unique solution? If so, what is the solution?

S3. The formula for the product of inertia of a plane area is

$$I_{xy} = \int_A xy \, dA$$

Using four Gauss points, determine the value of this integral for a unit square with the origin at the lower left corner. Check your answer with the exact formula.

S4. For each of Eqs. 9.52–9.58, define the sizes of all matrices indicated and make sure the matrix products are compatible.

S5. Write out in full the SF array for the three-node triangular element with one-point quadrature. Keep in mind that the element is the parent element defined in the *u-v* plane and that the quadrature point would be at the centroid of the element.

Numerical Experiments and Code Development

N1. Develop your own test problem for which the solution will be a linear function of *x* and/or *y*, and on some surfaces a flux is specified and a convective boundary condition is specified.

N2. Define a rectangular region for which our governing equation applies when

$$R_x = R_y = 100.0$$

$$B_x = B_y = 0.0$$

$$G = 0.0$$

$$H = 0.0$$

Select the element of your choice and limit the number to the minimum required to have at least one interior node. Specify that the upper and lower boundaries have zero flux. Run steady.m for each of the following four cases:

	Left Boundary Condition	Right Boundary Condition
NPBC	1	1
PHI	100	50
QS	0	0
HS	0	0
NPBC	1	0
PHI	100	0
QS	0	20
HS	0	0
NPBC	0	0
PHI	0	0
QS	20	90
HS	0	0
NPBC	0	0
PHI	0	70
QS	20	0
HS	0	10

For the last experiment, increase the value of HS on the right until the nodes on this surface approach the ambient temperature specified. Comment on each result and, when possible, compare it with the exact solution.

N3. Add to steady.m a section that will calculate the unknown right-hand-side values in $[SK]\{PHI\} = \{RHS\}$, after all PHI values have been determined.

Projects

P1. Conduct an analysis using steady.m and one of the higher-order elements, for any of the projects in Chapter 7.

P2. Of considerable importance in the evaluation of the thermal efficiency of homes is the heat loss through the basement walls and floor.

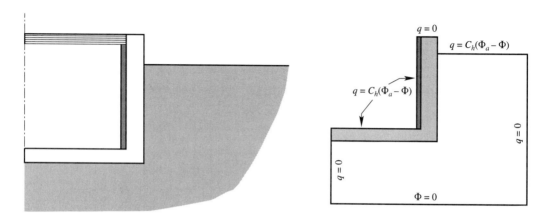

Consider the basement cross-section shown with the following dimensions, temperatures, and material properties:

Basement temperature	20°C
Outside air temperature	−18°C
Soil temperature at 12 m	0°C
Ceiling height	2.3 m
Room half width	4.6 m
Wall thickness	20 cm
Floor thickness	10 cm
Insulation thickness	5 cm
Depth of soil at bottom of floor slab	1.7 m
k for concrete walls and floor	1.5 W/m · °C
k for finished wall with insulation	0.03 J/s · m · °C
k for soil	0.03 J/s · m · °C
C_h for all surfaces	7.5 J/s · m² · °C

Assume the room is sufficiently long perpendicular to the page to allow a two dimensional analysis at its center. Also assume that a steady-state analysis will give a fairly accurate approximation as to the heat loss through the basement floor. With these assumptions, use the region of analysis and boundary conditions shown to determine the heat loss through the basement floor. Let the right-hand boundary be 10 meters from the outside wall and the lower boundary 12 meters below the soil surface.

P3. St. Venant's approach to the problem of torsion of shafts with noncircular cross-sections is to solve for the warping function, Ψ. The governing equation, which enforces equilibrium of the stresses is

$$\frac{\partial^2 \Psi}{\partial x^2} + \frac{\partial^2 \Psi}{\partial y^2} = 0$$

with boundary conditions

$$\left(\frac{\partial \Psi}{\partial x} - y\right)\frac{dy}{ds} - \left(\frac{\partial \Psi}{\partial y} - x\right)\frac{dx}{ds} = 0$$

which can be written

$$q = y\frac{dy}{ds} + x\frac{dx}{ds}$$

Here Ψ is a function only of the cross-sectional coordinates, x and y, with origin at the center of twist.

Use steady.m to solve for the warping function Ψ of a square shaft. For this case, the center of twist will be at the centroid.

Suggestion: In order to obtain this solution, it will be necessary to specify the q given above at every point on the boundary. This can be done in INITIAL.m but will require the calculation of dx/ds and dy/ds at that time. These quantities, however, are readily available in steady.m during the surface integration. Therefore, you might consider creating a new INCLUDE code, say BC.m, to be included in the surface integration.

For example, during the surface integration, change the code as follows:

steady.m

```
YK    =0.0;
XK    =0.0;
dxdu  =0.0;
dydu  =0.0;
PHIK  =0.0;
for L=1:NNPS;
  L1=NPSIDE(J,L);
  NPL=NP(I,L1);
  XK    =XK    +SF(4,L,K)*XORD(NPL);
  YK    =YK    +SF(4,L,K)*YORD(NPL);
  dxdu  =dxdu  +SF(5,L,K)*XORD(NPL);
  dydu  =dydu  +SF(5,L,K)*YORD(NPL);
  PHIK  =PHIK  +SF(4,L,K)*PHI(NPL);
end

dsdu = sqrt(dxdu^2 + dydu^2);
duds = 1/dsdu;

BC    <----------------------------------
```

BC.m

```
KEND=NUMQPT(2);
for K=1:KEND;
  HSK   =0.0;
  QSK   =0.0;
  for L=1:NNPS;
    L1=NPSIDE(J,L);
    NPL=NP(I,L1);
    QSK   =QSK   +SF(4,L,K)*QS(NPL);
    HSK   =HSK   +SF(4,L,K)*HS(NPL);
  end
end
```

This change would create a code that is the same as the current code. However, by revising BC.m you can calculate q by using the currently available dxdu and dydu as well as XK and YK. Note that dyds = dydu*duds, etc.

P4. One of the advantages of the finite element method is the ease with which it permits the user to set up and solve fairly complex problems, such as those involving coupled partial differential equations. Examples of coupled equations can be found in problems related to convective transport of thermal energy, pollutants, etc. As an example, consider the illustration, which shows fluid entering on the left with a uniform velocity and temperature. It flows around the pipe shown, which contains an energy source producing q units of energy per surface area of the pipe per unit time. Clearly, the temperature field will depend on the velocity field of the fluid. Also, the velocity field could depend on the temperature field due to a temperature-dependent viscosity. Thus, the two governing equations—for velocity and temperature—are coupled. You cannot solve one without information from the other.

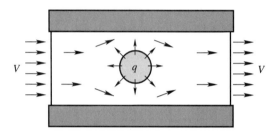

One method for solving such problems is to assume the first variable and solve for the second. With the solution for the second variable, then solve for the first variable. The new values for the first variable will now require another solution for the second variable, and so on. This procedure can be continued until there are only insignificant changes taking place in each iteration.

For this project, assume the fluid to be nonviscous and incompressible—hence, unaffected by the temperature field. Under these conditions there is no coupling of the flow equation with the temperature equation; thus, repeated iterations will not be needed. The flow can be completely determined by using the stream function approach discussed in Chapter 6.

Once it is determined, the temperature field can be determined by solving

$$\frac{\partial}{\partial x}\left(k\frac{\partial \Phi}{\partial x}\right) + \frac{\partial}{\partial y}\left(k\frac{\partial \Phi}{\partial y}\right) - \rho C_p u_x \frac{\partial \Phi}{\partial x} - \rho C_p u_y \frac{\partial \Phi}{\partial y} = 0.0$$

where

$$\Phi = \text{temperature}$$
$$k = \text{thermal conductivity of the fluid}$$
$$\rho C_p = \text{heat capacity of the fluid}$$
$$u_x = x \text{ component of velocity}$$
$$u_y = y \text{ component of velocity}$$

To obtain the velocities shown above, save the PHI values found from the flow analysis in a file, say VP. Then load these values in your INITIAL.m code when you run the temperature values. Having these values, the velocity components can be obtained using

$$u_x = \frac{\partial \Psi}{\partial y} = \lfloor DNDY \rfloor \{VP\}$$

and

$$u_y = -\frac{\partial \Psi}{\partial x} = \lfloor -DNDX \rfloor \{VP\}$$

For this project let $q = 1.0$ everywhere on the pipe's surface. Let the diameter of the pipe be 7.5 units of length and the region shown be 20 units by 50 units. Assume unit values for the thermal conductivity and the heat capacity of the fluid. Use the boundary conditions

At entrance	$\Phi = 0.0$
At exit	$q = 0.0$
On circular surface	$q = 1.0$

Because the equations are linear, the magnitude of the flow can be easily changed to any value desired by simply multiplying the VP array values by the appropriate factor. Compare temperature fields for entrance velocities corresponding to 0.0, 0.3, 0.6, and 1.

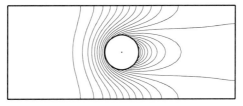

Temperature field for $V = 0.3$.

CHAPTER 10

ANALYSIS OF TRANSIENT BEHAVIOR

We now consider the analysis of problems where the dependent variable varies with time. Consider a function $\Phi(x, y, t)$ that satisfies

$$\frac{\partial}{\partial x}\left[k_x\frac{\partial \Phi}{\partial x}\right] + \frac{\partial}{\partial y}\left[k_y\frac{\partial \Phi}{\partial y}\right] = -Q + \mu\frac{\partial \Phi}{\partial t} \tag{10.1}$$

where Q, μ, k_x, and k_y can be functions of x, y, and t. The weak form of this equation begins with

$$\int_V \delta\Phi\left\{\frac{\partial}{\partial x}\left[k_x\frac{\partial \Phi}{\partial x}\right] + \frac{\partial}{\partial y}\left[k_y\frac{\partial \Phi}{\partial y}\right]\right\} dV - \int_V \delta\Phi\left\{-Q + \mu\frac{\partial \Phi}{\partial t}\right\} dV = 0.0 \tag{10.2}$$

and continues with the usual procedure of integration by parts:

$$\int_V \left\{\frac{\partial \delta\Phi}{\partial x}k_x\frac{\partial \Phi}{\partial x} + \frac{\partial \delta\Phi}{\partial y}k_y\frac{\partial \Phi}{\partial y}\right\} dV - \int_V \delta\Phi\left\{Q - \mu\frac{\partial \Phi}{\partial t}\right\} dV - \int_S \delta\Phi q\, dS = 0.0 \tag{10.3}$$

We derive our finite element approximation by considering Φ and $\partial\Phi/\partial t$ as two separate variables. Let

$$\dot{\Phi} = \frac{\partial \Phi}{\partial t} \tag{10.4}$$

and use the same finite element approximation for both Φ and $\dot{\Phi}$. Thus,

$$\Phi(x, y) = \lfloor N\rfloor\{\Phi\}$$
$$\dot{\Phi}(x, y) = \lfloor N\rfloor\{\dot{\Phi}\} \tag{10.5}$$

where $\dot{\Phi}$ represents the nodal point values for the derivative of Φ with respect to time. Note that we have two variables associated with each node, Φ and $\dot{\Phi}$. Substitution of these finite element approximations into the weak form gives, for a single element,

$$\delta J_e = \int_{V_e} \lfloor\delta\Phi\rfloor[N']^T [R][N']\{\Phi\}\, dV - \int_{V_e} \lfloor\delta\Phi\rfloor\{N\}\, Q\, dV$$

$$- \int_{S_s} \lfloor\delta\Phi\rfloor q\, dS + \int_{V_e} \lfloor\delta\Phi\rfloor\{N\}\,\mu\lfloor N\rfloor\{\dot{\Phi}\}\, dV \tag{10.6}$$

where we have used our usual notation. The first three integrals are familiar from previous chapters. However, the last integral is new. We define, therefore,

$$[C]_e = \int_{V_e} \{N\} \mu \lfloor N \rfloor dV \tag{10.7}$$

and write

$$\delta J_e = \lfloor \delta\Phi \rfloor [k]_e \{\Phi\}_e + \lfloor \delta\Phi \rfloor [C]_e \{\dot{\Phi}\}_e - \lfloor \delta\Phi \rfloor \{Q\}_e \tag{10.8}$$

which is the contribution to our weak form from element e. Summation of all such contributions from elements and surface segments gives us

$$\delta J = \lfloor \delta\Phi \rfloor [K] \{\Phi\} + \lfloor \delta\Phi \rfloor [C] \{\dot{\Phi}\} - \lfloor \delta\Phi \rfloor \{Q\} \tag{10.9}$$

For this to be zero for arbitrary $\delta\Phi$,

$$[K]\{\Phi\} + [C]\{\dot{\Phi}\} = \{Q\} \tag{10.10}$$

Note that the number of equations equals the number of values for either Φ or $\dot{\Phi}$, but not both. Hence, these equations govern the relationship between the two variables. If Φ were known, we could solve for $\dot{\Phi}$. On the other hand, if $\dot{\Phi}$ were known, we could solve for Φ. In previous chapters, we have assumed $\dot{\Phi}$ known (equal to zero) and solved for Φ.

10.1 FORWARD DIFFERENCE

If we are given the value of Φ at all nodes at time t, we can solve

$$[C]\{\dot{\Phi}\}_t = \{Q\}_t - [K]\{\Phi\}_t \tag{10.11}$$

for $\{\dot{\Phi}\}$ at time t. Once known, it is possible to use Eulerian integration to solve for $\{\Phi\}$ at time $t + \Delta t$; thus,

$$\{\Phi\}_{t+\Delta t} = \{\Phi\}_t + \{\dot{\Phi}\}_t \, \Delta t \tag{10.12}$$

We can combine Eqs. 10.11 and 10.12 by first solving Eq. 10.12 for $\{\dot{\Phi}\}$, to obtain

$$\{\dot{\Phi}\}_t = \frac{1}{\Delta t} \left[\{\Phi\}_{t+\Delta t} - \{\Phi\}_t \right] \tag{10.13}$$

and then substitute into Eq. 10.11 to obtain

$$[C]\frac{1}{\Delta t} \left[\{\Phi\}_{t+\Delta t} - \{\Phi\}_t \right] = \{Q\}_t - [K]\{\Phi\}_t \tag{10.14}$$

Rearrangement of Eq. 10.14 gives

$$\frac{1}{\Delta t}[C]\{\Phi\}_{t+\Delta t} = \{Q\}_t + \left[\frac{1}{\Delta t}[C] - [K]\right]\{\Phi\}_t \tag{10.15}$$

This method corresponds to the explicit method used in finite difference approximations, and, as in that formulation, it becomes unstable as Δt increases in value.

10.2 BACKWARD DIFFERENCE

The forward method allows us to calculate values of $\{\Phi\}$ at time $t + \Delta t$ at all interior points without any knowledge of the boundary conditions at this later time. Even without the problem of instability, which the method has, this characteristic might place some doubt on its desirability. However, both the error and the instability can be controlled by using small time steps. On the other hand, it is sometimes more efficient to use different methods that do not become unstable at larger time steps. One such method is the backward difference method.

We begin by writing our finite element equation (Eq. 10.10) for time $t + \Delta t$ rather than for time t. Thus,

$$[C]\{\dot{\Phi}\}_{t+\Delta t} = \{Q\}_{t+\Delta t} - [K]\{\Phi\}_{t+\Delta t} \tag{10.16}$$

Next, we consider our finite element expression for $\{\dot{\Phi}\}$ as a backward difference at time $t + \Delta t$ rather than a forward difference at time t. We therefore have

$$\{\dot{\Phi}\}_{t+\Delta t} = \frac{1}{\Delta t}\left\{\{\Phi\}_{t+\Delta t} - \{\Phi\}_t\right\} \tag{10.17}$$

which we substitute into Eq. 10.16 to obtain

$$\frac{1}{\Delta t}[C]\left\{\{\Phi\}_{t+\Delta t} - \{\Phi\}_t\right\} = \{Q\}_{t+\Delta t} - [K]\{\Phi\}_{t+\Delta t} \tag{10.18}$$

and write

$$\left[\frac{1}{\Delta t}[C] + [K]\right]\{\Phi\}_{t+\Delta t} = \{Q\}_{t+\Delta t} + \frac{1}{\Delta t}[C]\{\Phi\}_t \tag{10.19}$$

In this equation, all parameters used in the calculations of $[K]$ and $[C]$ should be evaluated at time $t + \Delta t$.

We now have a set of simultaneous equations for $\{\Phi\}_{t+\Delta t}$ that take into account the boundary conditions at $t + \Delta t$. In addition, this new formulation is stable for large time steps.

10.3 CENTRAL DIFFERENCE

We now have two formulations: the forward difference formulation,

$$\frac{1}{\Delta t}[C]\{\Phi\}_{t+\Delta t} = \{Q\}_t + \left[\frac{1}{\Delta t}[C] - [K]\right]\{\Phi\}_t \tag{10.20}$$

and the backward difference formulation,

$$\left[\frac{1}{\Delta t}[C] + [K] \right] \{\Phi\}_{t+\Delta t} = \{Q\}_{t+\Delta t} + \frac{1}{\Delta t}[C]\{\Phi\}_t \qquad (10.21)$$

For both of these formulations we used

$$\frac{1}{\Delta t} \left\{ \{\Phi\}_{t+\Delta t} - \{\Phi\}_t \right\}$$

as the approximation for $\{\dot{\Phi}\}$, whether at the beginning of the time step or at the end of the time step. However, this expression more accurately approximates $\{\dot{\Phi}\}$ at the center of the time step. Therefore, we write Eq. 10.10 as

$$[C]\{\dot{\Phi}\}_{t+\Delta t/2} = \{Q\}_{t+\Delta t/2} - [K]\{\Phi\}_{t+\Delta t/2} \qquad (10.22)$$

and use our difference expression interpreted as a central difference. For the right-hand side, we use the average value of $\{\Phi\}$ given by

$$\{\Phi\}_{t+\Delta t/2} \approx \frac{1}{2}\left\{ \{\Phi\}_t + \{\Phi\}_{t+\Delta t} \right\} \qquad (10.23)$$

Thus, Eq. 10.22 becomes

$$\frac{1}{\Delta t}[C]\left\{ \{\Phi\}_{t+\Delta t} - \{\Phi\}_t \right\} = \{Q\}_{t+\Delta t/2} - \frac{1}{2}[K]\left\{ \{\Phi\}_{t+\Delta t} + \{\Phi\}_t \right\} \qquad (10.24)$$

or

$$\left[\frac{1}{\Delta t}[C] + \frac{1}{2}[K] \right] \{\Phi\}_{t+\Delta t} = \{Q\}_{t+\Delta t/2} + \left[\frac{1}{\Delta t}[C] - \frac{1}{2}[K] \right] \{\Phi\}_t \qquad (10.25)$$

where now all parameters used in the calculations of $[K]$ and $[C]$ are evaluated at time $t + \Delta t/2$.

This is the central difference approximation and corresponds to the Crank-Nicolson method used in finite difference analyses. Although there are many other variations, the central difference formulation remains the most popular. It has a higher order of accuracy than either the forward difference formulation or the backward difference formulation, and it is also stable for large time steps. It is therefore widely used in finite element analyses.

The formulation used in most FEM codes is obtained by multiplying Eq. 10.25 by the time step Δt; thus,

$$[CPK]\{\Phi\}_{t+\Delta t} = \{\Delta Q\} + [CMK]\{\Phi\}_t \qquad (10.26)$$

where

$$[CPK] = [C] + \frac{\Delta t}{2}[K]$$

$$[CMK] = [C] - \frac{\Delta t}{2}[K]$$

$$\{\Delta Q\} = \Delta t\{Q\}_{t+\Delta t/2}$$

10.4 LUMPED-CAPACITANCE MATRIX

For all three methods discussed above, it is necessary to solve a set of simultaneous equations in each increment of the analysis. This can become very expensive in terms of computer time. When a similar approach is used with finite difference approximations, the forward difference method results in a set of equations that are decoupled. Hence, each member of $\{\dot{\Phi}\}$ can be solved for explicitly, without the necessity of solving a set of simultaneous algebraic equations. This is the reason that the forward difference approximation is referred to as the *explicit method* in finite difference literature, and why it is often preferred over the more accurate and stable Crank-Nicolson method.

In an effort to gain the same benefits from the forward difference method when finite element approximations are used, it is necessary that $[C]$ be a diagonal matrix. Several methods for diagonalizing $[C]$ are available. Before we discuss these methods, however, the physical interpretation of the approximation should be understood. If $[C]$ is a diagonal matrix, then the capacity of the material to store the quantity Φ will be concentrated at the nodes rather than distributed throughout the elements. Diagonalization, therefore, means that a distributed capacitance is replaced by many point capacitances. This is similar to replacing a distributed source with point sources. As long as the total lumped capacitance is the same as the total capacitance of the element, then changing the character of the material from one that has a continuously distributed capacitance to one that has many, closely placed point capacitances should not alter the gross behavior being modeled. We should therefore expect convergence when such an approximation is used. The term *lumped* is taken from the literature on the dynamics of structures, where a distributed mass is often lumped as point masses at the nodes. That procedure has long been referred to as the *lumped-mass method*. The full capacitance matrix is referred to as the *consistent-capacitance* matrix, because its formulation is consistent with the other terms in the weak form of the governing equation. Likewise, the term *consistent mass* is often used.

Lumping (or diagonalization) is done at the element level. Clearly, an assembly of diagonalized element matrices results in a diagonal global matrix. Therefore, we illustrate the methods employed using element matrices.

Row Summing. Row summing diagonalizes the matrix by replacing each diagonal with the sum of the terms on its row. Thus,

$$C(I, I) = \sum_{J=1}^{\text{NNPE}} \int_{V_e} N_I \mu N_J \, dV \tag{10.27}$$

$$C(I, J) = 0 \quad I \neq J$$

Because the shape functions always add to unity, this gives

$$C(I, I) = \int_{V_e} N_I \mu \, dV \tag{10.28}$$

We see that summing rows results in lumping the distributed capacitance at the nodes in exactly the same manner as we have lumped distributed sources at the nodes. We can therefore place a high degree of trust in this method to produce reliable results.

The following is a comparison of the consistent matrix with the lumped matrix for three parent-type elements. The numbers represent the fraction of the total capacitance of the element given by

$$c = \int_{V_e} \mu \, dV \tag{10.29}$$

Element	Consistent	Lumped
	$\dfrac{c}{12}\begin{bmatrix} 2 & 1 & 1 \\ 1 & 2 & 1 \\ 1 & 1 & 2 \end{bmatrix}$	$\dfrac{c}{3}\begin{bmatrix} 1 & 0 & 0 \\ 0 & 1 & 0 \\ 0 & 0 & 1 \end{bmatrix}$
	$\dfrac{c}{36}\begin{bmatrix} 4 & 2 & 1 & 2 \\ 2 & 4 & 2 & 1 \\ 1 & 2 & 4 & 2 \\ 2 & 1 & 2 & 4 \end{bmatrix}$	$\dfrac{c}{4}\begin{bmatrix} 1 & 0 & 0 & 0 \\ 0 & 1 & 0 & 0 \\ 0 & 0 & 1 & 0 \\ 0 & 0 & 0 & 1 \end{bmatrix}$
	$\dfrac{c}{180}\begin{bmatrix} 6 & 0 & -1 & -4 & -1 & 0 \\ 0 & 32 & 0 & 16 & -4 & 16 \\ -1 & 0 & 6 & 0 & -1 & -4 \\ -4 & 16 & 0 & 32 & 0 & 16 \\ -1 & -4 & -1 & 0 & 6 & 0 \\ 0 & 16 & -4 & 16 & 0 & 32 \end{bmatrix}$	$\dfrac{c}{3}\begin{bmatrix} 0 & 0 & 0 & 0 & 0 & 0 \\ 0 & 1 & 0 & 0 & 0 & 0 \\ 0 & 0 & 0 & 0 & 0 & 0 \\ 0 & 0 & 0 & 1 & 0 & 0 \\ 0 & 0 & 0 & 0 & 0 & 0 \\ 0 & 0 & 0 & 0 & 0 & 1 \end{bmatrix}$

In each case, the sum of the entries in each matrix equals unity; hence, the matrices represent the total capacitance of the element. For the three-node and four-node elements, the lumped form is very acceptable; however, the lumped form for the six-node element is not. The zero terms on the diagonal rule out the possibility of using this in an explicit formulation. Other higher-order elements have similar problems, with either zeros or negative values appearing on the diagonal. For this reason, the following method is preferable when high-order elements are used.

Scaled Diagonals. The scaled diagonals method diagonalizes the matrix by retaining the relative values of the diagonal terms but multiplying them by a scale factor to conserve the total capacitance of the matrix. Thus, the relative values of the diagonal terms in the lumped matrix are kept the same as they are in the consistent matrix. Because none of the terms is zero or negative (true also for higher-order elements), this method will always produce acceptable results.

For the three-node and four-node elements, the two methods of lumping give identical results. For the six-node element, the following matrix is obtained:

$$\frac{c}{114} \begin{bmatrix} 6 & 0 & 0 & 0 & 0 & 0 \\ 0 & 32 & 0 & 0 & 0 & 0 \\ 0 & 0 & 6 & 0 & 0 & 0 \\ 0 & 0 & 0 & 32 & 0 & 0 \\ 0 & 0 & 0 & 0 & 6 & 0 \\ 0 & 0 & 0 & 0 & 0 & 32 \end{bmatrix}$$

where, again,

$$c = \int_{V_e} \mu \, dV \tag{10.30}$$

In the above example, the elements illustrated are similar in shape to the parent element; hence, the values do not apply to all element shapes. However, the technique for diagonalizing the general element is exactly the same. In a FEM code, the full capacitance matrix should be formed. For row summing, each row can then be summed to create the diagonalized matrix. For scaled diagonals, the procedure is to sum all terms in the matrix. This gives the total capacitance for the element. Then sum the diagonals, which gives the capacitance associated with the diagonal terms. Once these two values are known, multiply each diagonal by the ratio of the element's total capacitance to that of the sum of the diagonals. The resulting diagonal terms will then be the diagonalized matrix.

10.5 STABILITY

When the lumped capacitance is used for the solution of a transient problem, the algorithm can become unstable. To understand this, consider the following one-dimensional problem:

$$k\frac{\partial^2 \Phi}{\partial x^2} = \mu \frac{\partial \Phi}{\partial t} \tag{10.31}$$

The weak form and finite element approximation create the following element matrices (see Chapter 5):

$$-\begin{bmatrix} +\dfrac{k}{\Delta x} & -\dfrac{k}{\Delta x} \\ -\dfrac{k}{\Delta x} & +\dfrac{k}{\Delta x} \end{bmatrix} \begin{bmatrix} \dfrac{\mu \, \Delta x}{3} & \dfrac{\mu \, \Delta x}{6} \\ \dfrac{\mu \, \Delta x}{6} & \dfrac{\mu \, \Delta x}{3} \end{bmatrix} \tag{10.32}$$

where Δx is the element length. The capacitance matrix on the right, when lumped, becomes

$$
\begin{bmatrix}
\dfrac{\mu\,\Delta x}{2} & 0 \\
0 & \dfrac{\mu\,\Delta x}{2}
\end{bmatrix}
\tag{10.33}
$$

If, for example, we have a mesh of four equal elements and constant k and μ, the assembled matrix equation will be

$$
\frac{k}{\mu(\Delta x)^2}
\begin{bmatrix}
-1 & 1 & 0 & 0 & 0 \\
1 & -2 & 1 & 0 & 0 \\
0 & 1 & -2 & 1 & 0 \\
0 & 0 & 1 & -2 & 1 \\
0 & 0 & 0 & 1 & -1
\end{bmatrix}
\begin{Bmatrix}
\Phi_1 \\
\Phi_2 \\
\Phi_3 \\
\Phi_4 \\
\Phi_5
\end{Bmatrix}
=
\begin{Bmatrix}
\frac{1}{2}\dot{\Phi}_1 \\
\dot{\Phi}_2 \\
\dot{\Phi}_3 \\
\dot{\Phi}_4 \\
\frac{1}{2}\dot{\Phi}_5
\end{Bmatrix}
\tag{10.34}
$$

For an interior node i, we can write the following equation:

$$
\frac{k\,\Delta t}{\mu(\Delta x)^2}
\begin{bmatrix} 1 & -2 & 1 \end{bmatrix}
\begin{Bmatrix}
\Phi_{i-1} \\
\Phi_i \\
\Phi_{i+1}
\end{Bmatrix}
= \dot{\Phi}_i \Delta t = \Phi_i(t + \Delta t) - \Phi_i(t)
\tag{10.35}
$$

Because the left-hand side of this equation is evaluated at time t, $\Phi_i(t)$ on the right can be moved to the left side to obtain

$$
\begin{bmatrix} r & 1 - 2r & r \end{bmatrix}
\begin{Bmatrix}
\Phi_{i-1} \\
\Phi_i \\
\Phi_{i+1}
\end{Bmatrix}
= \Phi_i(t + \Delta t)
\tag{10.36}
$$

where

$$r = \frac{k\Delta t}{\mu(\Delta x)^2} \qquad (10.37)$$

The cause of instability can now be illustrated by assuming that at time $t = 0$, all nodes have a Φ value of zero except node i, to which we give a small perturbation of magnitude ϵ. Then for the next increment of time, the above equation gives us the following values:

Time	Φ_{i-2}	Φ_{i-1}	Φ_i	Φ_{i+1}	Φ_{i+2}
0	0	0	ϵ	0	0
Δt	0	$r\epsilon$	$(1 - 2r)\epsilon$	$r\epsilon$	0

$$(10.38)$$

It is clear from this first step that if $r > \frac{1}{2}$, the sign of Φ_i will change, creating an oscillation in the results. However, if $r < 1$, the absolute value of $(1 - 2r)$ is less than 1 and the magnitude of Φ_i will have decreased even though its sign will have changed. Thus, we would expect continued integration to create oscillations of decreasing magnitude. Because the governing equation requires that such perturbations be dissipated and that the solution return to the steady-state conditions, our algorithm will approach the correct solution but with unwanted oscillations. On the other hand, for $r > 1$, the perturbation will not only change signs, it will increase in magnitude. Finally, for $r < \frac{1}{2}$, Φ_i will retain the same sign and decrease in magnitude. This final characteristic is the behavior we would prefer the algorithm to exhibit. In summary,

$\Delta t < \frac{1}{2}\mu(\Delta x)^2/k$ Stable with no oscillations

$\Delta t > \frac{1}{2}\mu(\Delta x)^2/k$ Stable, but oscillations may occur

$\Delta t > \mu(\Delta x)^2/k$ Unstable, oscillations with increasing magnitude may occur

In a two-dimensional finite element mesh, these conditions can be used as a guide for selecting Δt. The value to be used for the parameters k, μ, and Δx should be taken as the worst case found at any location in the mesh. Δx can be taken as the approximate diameter of a circle circumscribing the element at the location being considered.

10.6 PROGRAMMING PRELIMINARIES

The code we will use for the analysis of transient behavior is very similar to our other codes. It is necessary, however, to define certain new quantities associated with the transient analysis in INITIAL.m, as well as two new INCLUDE files: BC.m and MONITOR.m.

10.6.1 INITIAL.m. The user must initialize certain quantities in this INCLUDE file. These are

1. PHI: The initial values (i.e., the initial conditions) of the dependent variable must be specified at the beginning of the transient analysis.

2. MNI: This stands for the maximum number of increments.

3. DTIME: This is the increment of time to be used during the analysis, i.e., Δt.

4. INTsv: This stands for the interval to save results. Because a large number of time steps is normally taken during an analysis, it is unwise to save the results from each. This variable allows the user to specify how often the results should be saved. Note that if you were to specify MNI = 100,000 and wanted approximately 10 evenly spaced results to be saved, then you would specify

$$\text{INTsv} = 100{,}000/10 = 10{,}000$$

5. INCsv: This stands for increment to save results. Once the results are saved, this number is increased by the value set for INTsv. The user must specify the initial value for INCsv, which is usually either 1 or INTsv.

6. IRZ: This is the code for the type of analysis to be conducted:

$$\text{IRZ} = 0 \quad x\text{-}y \text{ coordinates}$$

$$= 1 \quad \text{Axisymmetric analysis about } x \text{ axis}$$

$$= 2 \quad \text{Axisymmetric analysis about } y \text{ axis}$$

You are free to calculate some of the quantities using variables you would rather specify. For example, you could specify the maximum time you wish to simulate and the time step, then calculate the maximum number of increments. Note that unlike other INITIAL.m codes, the boundary conditions are not specified at this time (see below).

10.6.2 BC.m. Transient behavior associated with many physical problems involves changing boundary conditions. In fact, these changing boundary conditions may be the root cause of the transient behavior. Thus, it is necessary to prescribe boundary conditions as a function of time and change them during the course of an analysis. We do this through the INCLUDE code BC.m. As with all INCLUDE codes, the user writes the code to match the particular problem being analyzed. It will be included in the main code at the beginning of each increment of time. Available for use in prescribing the boundary conditions will be the current values for time, the dependent variable, and the mesh data, including the NPcode values.

10.6.3 MONITOR.m. This INCLUDE code provides the user a way to monitor the transient analysis and make decisions during the analysis. It can be used for a variety of purposes. For example,

1. Save a specific nodal point value each increment to plot as a function of time.

2. Decide if the analysis has reached steady state. If so, set MNI equal to zero. The analysis will then stop.

3. Pause every so often to examine results.

This file can be left empty if the user so desires.

10.6.4 Output of Results. Some of the output will be sent to the screen, but the saved PHI values will be placed in a file named PHIp. The screen output shows which increments are saved in PHIp and the corresponding simulated time. The PHIp file can be used to plot the solution at any of the saved times. Each solution is saved as a record in PHIp; thus, if you wish to place the fifth record into the file PHI, you can simply write

$$PHI = PHIp(5,1:NUMNP)$$

and then program topo.m can be used to plot the solution. In addition, the last PHI values are saved in a separate file, PHIstrt. This file can be used as initial values for PHI if it is desired to restart the analysis where it left off.

10.7 PROGRAM transL.m

Program transL.m is the finite element code for the analysis of transient behavior using a lumped-capacitance matrix. In addition to the code, the following is required in the working directory:

Input Data	User's INCLUDEs	Supplied Functions
MESHo	INITIAL.m	SFquad.m
NODES	BC.m	
NP	COEF.m	
	MONITOR.m	

Because the capacitance matrix is lumped, there is no need for a Gaussian solution routine, nor is there a need to have new node numbers to reduce the bandwidth. The INPUT data files are the same as those used in previous codes to describe the mesh. The user INCLUDE codes will be described as part of the test problem at the end of this section.

Because row summing is used to diagonalize the capacity matrix, only three-node and four-node elements can be used with this code.

10.7.1 Flow Chart

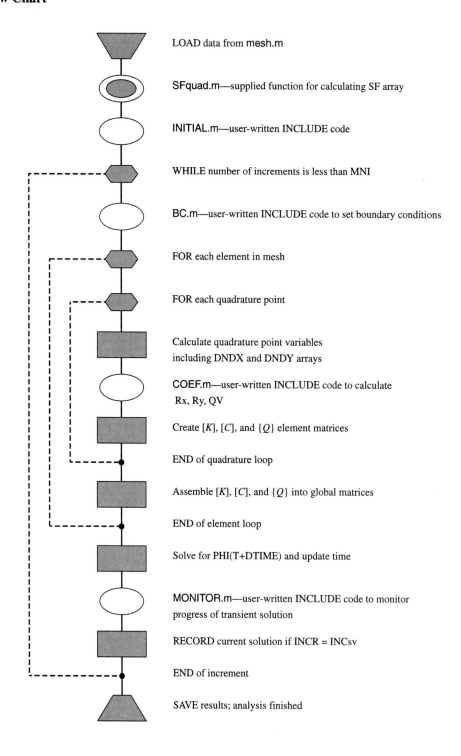

LOAD data from mesh.m

SFquad.m—supplied function for calculating SF array

INITIAL.m—user-written INCLUDE code

WHILE number of increments is less than MNI

BC.m—user-written INCLUDE code to set boundary conditions

FOR each element in mesh

FOR each quadrature point

Calculate quadrature point variables
including DNDX and DNDY arrays

COEF.m—user-written INCLUDE code to calculate
Rx, Ry, QV

Create $[K]$, $[C]$, and $\{Q\}$ element matrices

END of quadrature loop

Assemble $[K]$, $[C]$, and $\{Q\}$ into global matrices

END of element loop

Solve for PHI(T+DTIME) and update time

MONITOR.m—user-written INCLUDE code to monitor
progress of transient solution

RECORD current solution if INCR = INCsv

END of increment

SAVE results; analysis finished

10.7.2 Code

```
%-----------------------------------*
%                                   *
%         PROGRAM TransL            *
%                                   *
%         User must supply          *
%-----------------------------------*
% INITIAL.m  BC.m  COEF.m  MONITOR.m *
%-----------------------------------*

  clear
%----------------------------
% LOAD data from mesh
%----------------------------
 load MESHo    -ASCII
 load NODES    -ASCII
 load NP       -ASCII

%----------------------------------
%      General initialization of all
%      arrays and parameters
%----------------------------------
 NUMNP = MESHo(1);
 NUMEL = MESHo(2);
 NNPE  = MESHo(3);

 if NNPE ~= 3 & NNPE ~= 4
    error(...
    'NNPE must equal 3 or 4 in transL')
 end


 for I=1:NUMNP;
    XORD(I)=NODES(I,1);
    YORD(I)=NODES(I,2);
    NPcode(I)=NODES(I,3);
    NPBC(I)=0.0;
    PHIa(I)=0.0
 end

 NNPS=2;
 if NNPE       == 3
    NSPE=3;
 elseif NNPE == 4
        NSPE=4;
 end

 PI=3.1415926;
```

Transient analysis using lumped-capacitance matrix

$$\frac{\partial}{\partial x}\left(R_x\frac{\partial\Phi}{\partial x}\right) + \frac{\partial}{\partial y}\left(R_y\frac{\partial\Phi}{\partial y}\right) + Q = \mu\frac{\partial\Phi}{\partial t}$$

$$[C]\{\Phi\}_{t+dt} = \{Q\}_t + [CMK]\{\Phi\}_t$$

Load data to define mesh.

File	Contents
MESHo	NUMNP, NUMEL, NNPE
NODES	XORD, YORD, NPBC
NP	NP array

Variable	Definition
NUMNP	Number of nodal points
NUMEL	Number of elements
NNPE	Number of nodes per element

The lumped capacitance is produced by adding all terms in a row. This is valid only for three- and four-node elements.

Variable	Definition
XORD	Node's x coordinates
YORD	Node's y coordinates
NPcode	Nodal point code
NPBC	Nodal point boundary condition
PHIa	Nodal point ambient Φ value
NNPS	Number of nodes per side
NSPE	Number of sides per element

```
%----------------------------------------
% Create SF array
%----------------------------------------
   [SF,WT,NUMQPT,NPSIDE] = ShpFnctn(NNPE);
```

Create shape function array.

```
%----------------------------------------
% INCLUDE user's initialization
%----------------------------------------
   INITIAL
```

Include user-written INITIAL.m.

```
% ----------------------
% Begin transient analysis
% ----------------------
   disp(' ')
   disp('Segment    Time    ')
   disp('----------------- ')
```

Display column titles to screen for segment and corresponding time that are saved in PHIp.

```
   TIME=0.0;
   INCR=0;
   ISEG=0;
```

INCR = increment counter
ISEG = stored solution counter

```
   while  INCR <  MNI
```

Begin incremental solution.
MNI = maximum number of increments (specified in INITIAL.m)

```
     for I=1:NUMNP
        Q(I) =0.0;
        CL(I)=0.0;
     end
```

Initialize global arrays.
Q = right-hand side
CL = lumped-capacitance matrix

```
%    ----------------------------
%    INCLUDE user-written BC.m
%    to set all boundary conditions
%    ----------------------------
     BC
```

Include user-written BC.m to set all boundary conditions for current increment.

```
%    ----------------------------
%    Formation of C and S matrices
%    ----------------------------
     for I=1:NUMEL
       LMNT = I;
       for J=1:NNPE
         QE(J)=0.0;
         C(J)=0.0;
         for K=1:NNPE
            S(J,K)=0.0;
         end
       end
```

Begin compiling global matrices element by element.

Initialize current element matrices for Gaussian quadrature.

```
%   ----------------------
%   Begin volume quadrature
%   ----------------------
    JEND=NUMQPT(1);
      for J=1:JEND

        RJAC(1,1)=0.0;
        RJAC(1,2)=0.0;
        RJAC(2,1)=0.0;
        RJAC(2,2)=0.0;

        XJ=0.0;
        YJ=0.0;
        PHIJ=0.0;

        for K=1:NNPE
          NPK=NP(I,K);
          RJAC(1,1)=RJAC(1,1)+...
              SF(2,K,J)*XORD(NPK);
          RJAC(1,2)=RJAC(1,2)+...
              SF(3,K,J)*XORD(NPK);
          RJAC(2,1)=RJAC(2,1)+...
              SF(2,K,J)*YORD(NPK);
          RJAC(2,2)=RJAC(2,2)+...
              SF(3,K,J)*YORD(NPK);
          XJ=XJ+SF(1,K,J)*XORD(NPK);
          YJ=YJ+SF(1,K,J)*YORD(NPK);
          PHIJ=PHIJ+SF(1,K,J)*PHI(NPK);
        end

        DETJ=RJAC(1,1)*RJAC(2,2)...
            -RJAC(2,1)*RJAC(1,2);
        if DETJ <= 0
          error(' DETJ <= 0 in TRANSL.m')
        end

        DV=DETJ;
        if IRZ == 1
          DV=2.0*PI*YJ*DV;
        elseif IRZ == 2
          DV=2.0*PI*XJ*DV;
        end

        RJACI(1,1)=+RJAC(2,2)/DETJ;
        RJACI(1,2)=-RJAC(1,2)/DETJ;
        RJACI(2,1)=-RJAC(2,1)/DETJ;
        RJACI(2,2)=+RJAC(1,1)/DETJ;
```

Begin Gaussian quadrature.

Initialize Jacobian matrix.

Initialize quadrature coordinates.

$$\left[J \frac{(x,y)}{(u,v)} \right] = \begin{bmatrix} \dfrac{\partial x}{\partial u} & \dfrac{\partial x}{\partial v} \\[2ex] \dfrac{\partial y}{\partial u} & \dfrac{\partial y}{\partial v} \end{bmatrix}$$

$$= \begin{bmatrix} \left\lfloor \dfrac{\partial N}{\partial u} \right\rfloor \{X\} & \left\lfloor \dfrac{\partial N}{\partial v} \right\rfloor \{X\} \\[2ex] \left\lfloor \dfrac{\partial N}{\partial u} \right\rfloor \{Y\} & \left\lfloor \dfrac{\partial N}{\partial v} \right\rfloor \{Y\} \end{bmatrix}$$

$$x = \lfloor N \rfloor \{X\}$$

$$y = \lfloor N \rfloor \{Y\}$$

$$\text{DETJ} = \left\| \begin{matrix} \dfrac{\partial x}{\partial u} & \dfrac{\partial x}{\partial v} \\[2ex] \dfrac{\partial y}{\partial u} & \dfrac{\partial y}{\partial v} \end{matrix} \right\|$$

Check sign of $\|J\|$ to determine if element is correctly defined.

Account for axisymmetric analysis if specified.

$$\begin{bmatrix} \dfrac{\partial u}{\partial x} & \dfrac{\partial u}{\partial y} \\[2ex] \dfrac{\partial v}{\partial x} & \dfrac{\partial v}{\partial y} \end{bmatrix} = \begin{bmatrix} \dfrac{\partial x}{\partial u} & \dfrac{\partial x}{\partial v} \\[2ex] \dfrac{\partial y}{\partial u} & \dfrac{\partial y}{\partial v} \end{bmatrix}^{-1}$$

```
        for K=1:NNPE
            DNDX(K)=RJACI(1,1)*SF(2,K,J)...
                    +RJACI(2,1)*SF(3,K,J);
            DNDY(K)=RJACI(1,2)*SF(2,K,J)...
                    +RJACI(2,2)*SF(3,K,J);
        end

%       -------------------------------
%       INCLUDE user's COEF.m
%       -------------------------------
        COEF

%       -------------------------------
%       Create element matrices
%       -------------------------------
        for K=1:NNPE
            C(K)=C(K) + ...
                WT(1,J)*SF(1,K,J)*RC*DV;
            QE(K)=QE(K)+...
                WT(1,J)*SF(1,K,J)*QV*DV;
            for L=1:NNPE
                S(K,L)=S(K,L)+WT(1,J)*...
                    (DNDX(K)*RX*DNDX(L) ...
                    +DNDY(K)*RY*DNDY(L))*DV;
            end
        end
    end
%   --------   end of volume quadrature

%   -----------------------
%   Begin surface quadrature
%   -----------------------
    for J=1:NSPE
        Hchk=1;
        Qchk=1;
        for K=1:NNPS;
            J1=NP(I,NPSIDE(J,K));
            Hchk=Hchk*HS(J1);
            Qchk=Qchk*QS(J1);
        end

        if Hchk ~= 0 | Qchk ~= 0
            KEND=NUMQPT(2);
            for K=1:KEND
```

$$\begin{bmatrix} \dfrac{\partial N}{\partial x} \\[2mm] \dfrac{\partial N}{\partial y} \end{bmatrix} = \begin{bmatrix} \dfrac{\partial u}{\partial x} & \dfrac{\partial v}{\partial x} \\[2mm] \dfrac{\partial u}{\partial y} & \dfrac{\partial v}{\partial y} \end{bmatrix} \begin{bmatrix} \dfrac{\partial N}{\partial u} \\[2mm] \dfrac{\partial N}{\partial v} \end{bmatrix}$$

Include user's **COEF**.m to define

$$RX = R_x$$
$$RY = R_y$$
$$QV = Q$$
$$RC = \mu$$

$$\{C\}_e = \int \{N\}\mu \, dV$$

$$\{Q\}_e = \int \{N\}Q \, dV$$

$$[S]_e = \int [N']^T [R] [N'] dV$$

Examine each side of current element.

If all nodes on a given side have nonzero values for either HS or QS, then a surface integration is necessary.

This is a check to determine if there are zero values of these quantities.

If either Hchk or Qchk is nonzero, then begin surface quadrature.

NUMQPT(2) = number of quadrature points for surface integrations

```
        XK    =0.0;
        YK    =0.0;
        DXDu  =0.0;
        DYDu  =0.0;
        HSK   =0.0;
        QSK   =0.0;
        PHIK  =0.0;
        for L=1:NNPS
            L1=NPSIDE(J,L);
            NPL=NP(I,L1);
            XL=XORD(NPL);
            YL=YORD(NPL);
            QSL=QS(NPL);
            HSL=HS(NPL);
            PHIL=PHIa(NPL);

            XK    =XK    +SF(4,L,K)*XL;
            YK    =YK    +SF(4,L,K)*YL;
            DXDu  =DXDu  +SF(5,L,K)*XL;
            DYDu  =DYDu  +SF(5,L,K)*YL;
            QSK   =QSK   +SF(4,L,K)*QSL;
            HSK   =HSK   +SF(4,L,K)*HSL;
            PHIK  =PHIK  +SF(4,L,K)*PHIL;
        end
        DS=sqrt(DXDu^2+DYDu^2);
        if IRZ == 1
            DS=2.0*PI*YK*DS;
        end
        if IRZ == 2
            DS=2.0*PI*XK*DS;
        end

    for L=1:NNPS
        L1=NPSIDE(J,L);
        if Qchk ~= 0;
            QE(L1)=QE(L1)+WT(2,K)...
                    *SF(4,L,K)*QSK*DS;
        end

        if Hchk ~= 0;
            QE(L1)=QE(L1)+WT(2,K)...
                    *SF(4,L,K)*HSK*PHIK*DS;
            for M=1:NNPS
                M1=NPSIDE(J,M);
                S(L1,M1)=S(L1,M1) ...
                    +WT(2,K)*SF(4,L,K)*...
                        HSK*SF(4,M,K)*DS;
            end
        end
    end
end
```

Use nodal point values to interpolate parameters at current quadrature point, point K.

Variable	Definition
XK	x coordinate
YK	y coordinate
DXDu	dx/du
DYDu	dy/du
QSK	Specified surface flux, q_s
HSK	h in $q_s = h(\Phi_\infty - \Phi)$
PHIK	Φ_∞ in $q_s = h(\Phi_\infty - \Phi)$

Determine ratio of ds on surface to du along parent segment. Adjust, if appropriate, for axisymmetric analysis.

If Qchk is nonzero, then integrate

$$\int_s \{N\} q_s \, ds$$

If Hchk is nonzero, then integrate

$$\int_s \{N\} h \Phi_\infty \, ds$$

and

$$\int_s \{N\} h \lfloor N \rfloor \, ds$$

```
            end % surface quadrature
          end % Hchk and Qchk
        end % search through element sides


%    ----------------------
%    Create global vectors
%    ----------------------
    for J=1:NNPE
        NPJ=NP(I,J);
        Q(NPJ)=Q(NPJ)+QE(J);
        CL(NPJ)=CL(NPJ)+C(J);
        for K=1:NNPE
            NPK=NP(I,K);
            Q(NPJ)=Q(NPJ)-S(J,K)*PHI(NPK);
        end
    end

end % loop through elements

%    --------------------------------
%    Increment variables to next time
%    --------------------------------
    for I=1:NUMNP
        dPdt(I) = 0;
        if NPBC(I) ~= 1
            dPdt(I) = Q(I)/CL(I);
            PHI(I)=PHI(I)+dPdt(I)*DTIME;
        end
    end

MONITOR

TIME=TIME+DTIME;
INCR=INCR+1;
if INCR >= MNI
    INCsv=INCR;
end


if INCR >= INCsv
    INCsv=INCsv+INTsv;
    ISEG=ISEG+1;
    for I=1:NUMNP
        PHIp(ISEG,I)=PHI(I);
    end
```

Finished with volume and surface integrations for current element.

Add element integrations to global quantities.

$$[C]\{\dot{\Phi}\} = \{Q\} - [S]\{\Phi\}$$

Note: $[C]$ is a diagonal matrix and $\{CL\}$ are the lumped values.

The global integration has now been completed.

Solve dPdt = $\dot{\Phi}$ at each node.

$\dot{\Phi}$ equals zero at all nodes for which Φ has been specified. Otherwise, increment the current Φ value.

$$\Phi(t + \Delta t) = \Phi(t) + \dot{\Phi}\Delta t$$

Include user's **MONITOR**.m to monitor the integration.

Increment time and increment counter.

If this is last increment, save results.

Save results if INCR = INCRsv.

Increment marker for next increment to save.
Increment counter of saved results.

Save results in **PHIp**.

```
    a = [ISEG TIME ];
    fprintf...
     (1,'%4i  %10.2e  \n',a)
  end
end % of transient analysis




save PHIp      PHIp   -ASCII
save PHIstrt PHI    -ASCII




fprintf(1,'\n NUMNP = %4i \n',NUMNP)

disp('                                  ')
disp(' Nodal point values for the       ')
disp(' dependent variable, PHI,         ')
disp(' corresponding to the above       ')
disp(' segments at the times shown      ')
disp(' are recorded on file PHIp in     ')
disp(' your working directory.  To      ')
disp(' extract any segment, e.g. 5,     ')
disp(' for plotting or other uses, type:')
disp('                                  ')
disp('        load PHIp                 ')
disp('        PHI = PHIp(5,1:N)         ')
disp('        save PHI PHI -ASCII       ')
disp('                                  ')
disp(' where N is the numerical value   ')
disp(' for NUMNP shown above.           ')
disp('                                  ')
```

Report to user:

Record No.   Time

Save results in ASCII files.

PHIstrt = last record in PHIp—can be used as initial values of PHI if analysis is to be continued.

Tell user where results are located and how to access them.

10.7.3 Test Problem. A test problem for transL.m can be developed by defining a region for which the steady-state solution is found exactly by the finite element approximation, and then perturbing that solution by a small amount to determine if transL.m drives the perturbed values back to the steady-state values.

As an example, consider the following mesh covering a region for which

$$\frac{\partial}{\partial x}\left(3\frac{\partial \Phi}{\partial x}\right) + \frac{\partial}{\partial y}\left(3\frac{\partial \Phi}{\partial y}\right) = 2\frac{\partial \Phi}{\partial t}$$

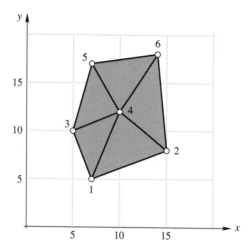

Let the steady-state solution for the problem be

$$\Phi = 3x - 5y$$

and specify values at all boundary points to agree with this solution. If the interior node is given any value other than that specified by the steady-state solution, then transL.m should force the change in that value to move toward the steady-state value with time.

The following data and INCLUDE files correspond to this problem and can be used as a test problem for transL.m.

DATA Files

MESHo

```
6   %   NUMNP
5   %   NUMEL
3   %   NNPE
```

NODES

```
%========================
%   x      y    np-code
%========================
    7      5      1
   15      8      1
    5     10      1
   10     12      0    % Interior node
    7     17      1
   14     18      1
```

NP

```
%==========================
%   Nodes for each element
%==========================
     1    2    4
     1    4    3
     4    2    6
     3    4    5
     4    6    5
```

INCLUDE Files

COEF.m

```
%----------------------------------------------
%
% d       d@      d       d@            d@
% -- (RX -- ) + -- (RY -- ) = RC --    - QV
% dX      dX      dY      dY            dT
%
%    Must specify RX, RY, RC, and QV
%----------------------------------------------

         RX = 3;
         RY = 3;
         RC = 2;
         QV = 0;
```

MONITOR.m

```
%-------------------------------------------
%               MONITOR.m
%
%    Monitor transient results.
%    Currently at increment = INCR+1
%-------------------------------------------


% -----------------------------------------
% Save values of PHI(4) for plotting
% -----------------------------------------
  PHI4(INCR+1) = PHI(4);

% -----------------------------------------
% Calculate difference between current
% PHI and the steady-state value.
% -----------------------------------------
E = PHI(4) - (3*XORD(4) -5*YORD(4));
if abs(E) <= 0.0001
    MNI = 0       % Turn off solution
end
```

INITIAL.m

```
%-------------------------------------
%              INITIAL.m
%  Must specify:
%-------------------------------------
%   PHI      Initial values for PHI
%   MNI      Maximum number of increments
%   DTIME    Time step
%   INCsv    First increment to save PHI
%            values
%   INTsv    Subsequent intervals to save
%            PHI values
%   IRZ      Axi-symmetric designation
%-------------------------------------

%  Initialize all PHI values
%-------------------------------------
    for I=1:NUMNP
        PHI(I)  = 3*XORD(I) -5*YORD(I);
    end
    PHI(4)=0.0;  % Perturb node 4


%-------------------------------------
%     Calculate time step.
%
%   R=3, mu=2, dx = 5
%
%     dT = 0.5*(mu)*(dx^2)/R
%     dT = 0.5*(2)*(5^2)/3  = 8.34
%-------------------------------------
 DTIME = 11.0;    % Unstable
 DTIME = 10.0;    % Stable but
                  % severe oscillation
 DTIME = 8.00;    % Stable but
                  % some oscillation
 DTIME = 4.00;    % Stable


%-------------------------------------
% Calculate max. number of increments
%-------------------------------------
 MNI   = round(900/DTIME)

 IRZ   = 0;    % x-y coordinates
 INCsv = 1;    % First increment to
               % save PHI values
 INTsv = 1;    % Subsequent intervals
               % to save PHI values
```

BC.m

```
%-------------------------------------
%              BC.m
%
%  For each increment of the analysis,
%  the following values must be specified
%  at each node:
%
%  PHIa = ambient value of PHI
%    Q  = specified point source
%    QV = specified distributed source
%         over volume
%    QS = specified flux on surface
%    HS = coefficient for surface
%         convection
%  NPBC = nodal point boundary condition
%
%   PHI = current value of PHI.  Need
%         not be specified if it remains
%         constant.
%-------------------------------------

for I=1:NUMNP
 code=NPcode(I);

  if code == 0      % node not on boundary
    PHIa(I) = 0.;
    Q(I)    = 0.;
    QV(I)   = 0.;
    QS(I)   = 0.;
    HS(I)   = 0.;
    NPBC(I) = 0;

  elseif code == 1  % point is on boundary
    PHIa(I) = 0;;     % ambient PHI
    Q(I)    = 0.;     % point source
    QV(I)   = 0.;     % distributed source
    QS(I)   = 0;      % surface flux
    HS(I)   = 0;      % coefficient of
                      % surface convection
    NPBC(I) = 1;      % Phi is known

    PHI(I)  = 3*XORD(I) -5*YORD(I);
                      % steady-state value
                      % on boundary
  end
end
```

10.7.4 Suggestions for Conducting an Analysis

1. Begin a transient analysis by running only a few increments (e.g., 5). By examining results from such an analysis, it is often easy to spot errors made in assigning boundary conditions or in assigning values for the coefficients.

2. For most problems, it is not necessary to save results for each increment. Use INTsv for this purpose. If you wish to save, for example, 10 increments for plotting, then in your INITIAL.m file, specify

$$INTsv = MNI/10;$$

3. For problems that will require many increments of time, divide your analysis into smaller segments of time and use the PHIstrt to begin the next segment. You might want to add to your INITIAL.m file the option for reading this start-up file, for example,

```
disp(' ')
disp(' Do you have a start up file?')
disp(' ------------------------------')
disp(' ENTER 1 if yes. ')
disp(' ------------------------------')
ans = input(' < ');
if ans == 1
    load PHIstrt -ASCII
    for I =1:NUMNP
        PHI(I)=PHIstrt(I);
    end
end
```

4. Be creative with your MONITOR.m files. If your analysis approaches a steady-state solution, monitor changes in your PHI values and stop the analysis when the changes become insignificant in terms of what you are interested in obtaining. This was done in the test problem. Note that changing MNI to zero will automatically shut down the analysis with the current increment.

    It is sometimes useful to keep track of a particular PHI value with time, as was done with PHI(4) in the test problem. You can then plot the result after the analysis to determine the rate of convergence. For the test problem, after each analysis simply type

```
plot(PHI4)
```

You might also want to save the maximum and minimum values of PHI at any instant of time and plot them for comparison purposes. This is suggested in one of the projects at the end of this chapter as a method of determining how close an object is to having a uniform temperature.

EXERCISES

Study Problems

S1. Perform a dimensional analysis of Eq. 10.1 when it represents the heat equation.

S2. For a convective-type boundary condition, explain why specifying

$$QS(I)=HS*(PHIa - PHI(I))$$

is the same as specifying

$$HS(I) \quad and \quad PHIa(I)$$

in BC.m for a forward difference approximation, but not for a backward or central difference approximation.

S3. Linear, second-order, partial differential equations are classified as being either elliptic, parabolic, or hyperbolic. To which classification does our equation for transient behavior belong? What is the characteristic direction associated with our equation? Does this have any application to our forward, central, and backward difference approaches to our problem?

S4. Shown in the following figure is a grid in the x-t plane for the analysis of a one-dimensional heat transfer problem. Two nodal point specifications are shown: a coarse grid represented by the filled circles, and a finer grid represented by both the filled and empty circles. The horizontal lines are the times corresponding to the given time step. If the temperature is known at time zero but the boundary conditions are unknown for $t > 0$, show that a lumped-capacitance algorithm will nevertheless predict the temperatures at all nodes that fall on or beneath the lower triangle when using the coarse grid, and all nodes that fall on or beneath the upper triangle when using the finer grid.

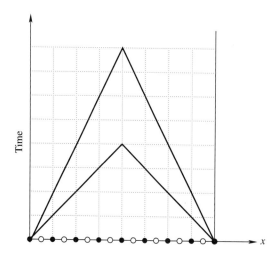

From this, determine how far into time it is possible to predict the temperature at the center of the rod for a mesh of N elements using the maximum time step possible for no oscillations to occur. Write your answer in terms of $\mu L^2/k$.

The preceding graphical analysis indicates that for a given time step, it is possible to determine the temperature at the center of the rod as far into the future as we desire, with no consideration of boundary conditions, simply by taking smaller and smaller elements. Of course, this is nonsense and is one reason why the forward difference method must be used with care. Explain why this is not a problem with use of either the backward or central difference algorithms.

S5. Consider the spring-dashpot system shown, and let the unstretched length of the spring correspond to $x = 0$. If inertial effects are negligible, then for any displacement x, equilibrium demands

$$kx = -\mu \frac{dx}{dt}$$

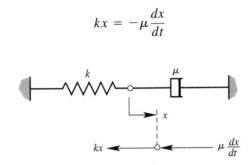

(a) Write the forward difference approximation for $x(t + \Delta t)$.
(b) Write the backward difference approximation for $x(t + \Delta t)$.
(c) Write the central difference approximation for $x(t + \Delta t)$.
(d) For each of the preceding approximations, determine the ranges for Δt for
    i. Stable with no oscillations
    ii. Stable but with oscillations
    iii. Unstable
(e) Write a simple computer code to run both forward and backward difference approximations for this problem. For a unit of time equal to μ/k, run steps of 0.01, 0.5, 1.9, 2.1, and 2.5 and plot the results. Use a total time of 5.0 units. Compare each with the exact solution. Run the backward difference algorithm with time steps of 1.5 and 2.05 for 50 units of time. Comment on all results.
(f) Sketch the exact solution of the problem for some initial displacement on a graph of x versus t. Mark off a large enough Δt from $t = 0$ so that at that time, the slope for the exact solution would be fairly small. Then sketch on this graph each slope used for the three methods to predict $x(t + \Delta t)$ and observe what each predicts. From these observations, convince yourself that both the backward difference and the central difference methods are unconditionally stable, that is, stable no matter how large a Δt is used. Next, convince yourself that the central difference is the most accurate of the three methods. Finally, graphically determine on the graph values for Δt that would, for the forward difference method, correspond to
    i. Stable with no oscillations
    ii. Stable with oscillations
    iii. Unstable

Numerical Experiments and Code Development

N1. Create a test problem using four-node elements. Calculate the estimated time step for stability as given by the one-dimensional analysis and numerically determine how close it is to being correct.

N2. Convert transL.m to transC.m for a consistent-capacitance matrix and use the central difference method.

N3. Change the method of lumping in transL.m from row summing to scaled diagonals. Compare results for the two methods using a test problem with three- and four-node elements. Then test the code using six-node elements.

N4. Revise transL.m to solve the more general equation

$$\frac{\partial}{\partial x}\left[R_x\frac{\partial\Phi}{\partial x}\right] + \frac{\partial}{\partial y}\left[R_y\frac{\partial\Phi}{\partial y}\right] + B_x\frac{\partial\Phi}{\partial x} + B_y\frac{\partial\Phi}{\partial y} + G\Phi = -Q + \mu\frac{\partial\Phi}{\partial t}$$

You may wish to use as your guide program steady.m, or even consider revising it rather than transL.m.

N5. Write a code for the analysis of the one-dimensional problem

$$\rho C_p\frac{\partial\Phi}{\partial t} = \frac{\partial}{\partial x}\left[k\frac{\partial\Phi}{\partial x}\right] + Q$$

for both lumped-capacitance and consistent-capacitance $[C]$ matrices. Use this program to study the difference between the two approaches as applied to a rod of unit length, having unit values of ρC_p and k, when one end is suddenly heated to 100 degrees above its initial temperature. Of the two approaches, which one is more stable as Δt is increased? Use the forward difference algorithm for both.

N6. The equation for the free vibration of a tightly stretched wire (e.g., in a piano or on a guitar) is

$$T\frac{\partial^2 y}{\partial x^2} = \rho\frac{\partial^2 y}{\partial t^2}$$

where the sign convention and notation are the same as that given in Chapters 1–3, with the exception that y is now both a function of x and t, and ρ is the mass per unit length of the wire. Revise program wire.m to solve for the time-dependent positions of the wire as given by the above equation. Here are some suggestions:

(a) Create arrays for $\{Y\}$ and $\{\dot{Y}\}$, and set their initial values in a user INCLUDE statement.

(b) Solve for nodal values of $\ddot{y}$, using the form

$$[M]\{\ddot{Y}\} = [K]\{Y\} - \{F\}$$

(c) Use an expansion of Eulerian integration for second-order equations, given by

$$\left\{\frac{\partial y}{\partial t}\right\}_{t+\Delta t} = \left\{\frac{\partial y}{\partial t}\right\}_t + \left\{\frac{\partial^2 y}{\partial t^2}\right\}_t \Delta t$$

$$\{y\}_{t+\Delta t} = \{y\}_t + \left\{\frac{\partial y}{\partial t}\right\}_t \Delta t$$

(d) In the above formula, let $[M]$ be the lumped-mass matrix (a diagonalized matrix) with the mass at each node assigned to be equal to the sum of half the masses of the elements on each of its sides.

(e) Store $[K]$ as a nonsymmetric matrix (three columns) for the multiplication of $[K]\{Y\}$.

(f) After assembly of $[K]$, $\{M\}$, and $\{F\}$, write a short loop to solve the preceding equations for as many time steps as desired.

Test the code by specifying that the wire is released from rest in the position

$$y(t = 0) = \sin(2\pi x/L)$$

Use unit length, unit tension, unit density, 21 evenly spaced nodes, and $\Delta t = 0.00001$. Run the analysis for 55,000 increments and save the results every 7000 increments, to obtain

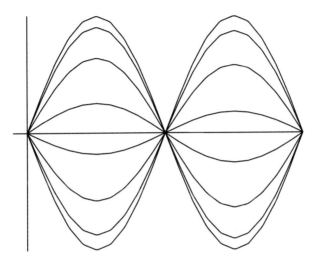

Projects

P1. **Cooling after shrink fit.** A common method used to assemble two parts during a manufacturing process is known as *shrink fitting*. As an example, consider the cylindrical rod and shield shown. The rod is to be fitted into the shield by heating the shield to 85°C and cooling the rod to −10°C. This will make the fit sufficiently tight to prevent slipping during periods of predicted temperature changes in actual use.

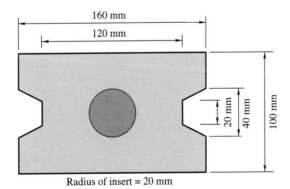

Radius of insert = 20 mm

The rod is copper and the shield is aluminum; they have the following physical properties:

Material	k [N·m/°C·m·s]	ρ [kg/m³]	C_p [N·m/kg·°C]
Aluminum	240	2700	917
Copper	397	8900	384

After assembly, the part is allowed to cool in air at a temperature of 15°C. The coefficient of convective heat transfer for all outside surfaces is equal to 50 N·m/°C·m²·s. Because the dimension of the part perpendicular to the page is large compared to the other dimensions, a two-dimensional analysis to determine the temperature in the part as a function of time is appropriate.

Important in the design of the manufacturing process is the time necessary for all parts of the rod to be less than 75°C. Use transL.m, with $k_x = k_y = k$, $Q = 0$, and $\mu = \rho C_p$, to determine this time. For the analysis, make use of symmetry in both the horizontal and vertical directions. During the analysis, monitor both the maximum and minimum temperatures and plot them as functions of time.

P2. Thermal history related to the heat of hydration. A problem of interest to engineers during the construction of concrete dams and other large concrete structures is the thermal history created by the heat of hydration of the concrete mix. This history is important due to its effect on the curing of the concrete as well as possible thermal stresses that it can create. The accompanying figure shows the cross-section of a concrete dam to be constructed at a rate of 4 ft/day. Use transL.m to determine the temperature in the dam during construction and for 120 days after the completion of construction. Use MONITOR.m to record the largest temperature anywhere in the dam at each time step. Plot this temperature with time.

The governing equation for heat transfer can be written as

$$\frac{\partial}{\partial x}\left[R\frac{\partial \Theta}{\partial x}\right] + \frac{\partial}{\partial y}\left[R\frac{\partial \Theta}{\partial y}\right] = -G + \frac{\partial \Theta}{\partial t}$$

with the boundary condition $q = H(\Theta_A - \Theta)$. Here

$$R = \frac{k}{\rho C_p}[\text{ft}^2/\text{day}] \qquad G = \frac{Q}{\rho C_p}[\text{°F/day}] \qquad H = \frac{h}{\rho C_p}[\text{°F/day}]$$

where

$R = 1.44$
$G = 4.5 \quad$ for $0 < t < 4$ days
$\quad = 2.5 \quad$ for $4 < t < 8$ days
$\quad = 0.3 \quad$ for $8 < t < 16$ days
$\quad = 0.0 \quad$ for $16 < t$
$H = \quad 4.0 \quad$ for air
$\quad = 10.0 \quad$ for foundation rock
$\Theta_A = 70\text{°F} \quad$ for air
$\quad = 50\text{°F} \quad$ for foundation rock
$\Theta_o = 70\text{°F}$ (temperature of concrete when poured)

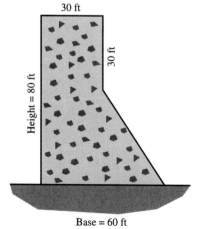

30 ft

30 ft

Height = 80 ft

Base = 60 ft

Comments: Problems that involve moving boundaries such as we have here can be analyzed by adding and/or removing new elements as needed, or by creating a completely new mesh when needed. This, of course, creates certain logistic problems in the programming. An easier way is to begin with a mesh for the completed dam, and for elements in an area of the dam not yet completed, consider that they represent air or a pseudo-material that simulates the boundary condition desired.

For this particular problem, the new concrete is being added at ambient temperature; hence, it is reasonable to assume that there is no heat transfer between the newly added concrete and the air above it. This can be simulated by assigning a low value for k. Thus, very little heat will be transferred across the simulated surface, and the nodes above this surface will remain at a constant temperature. If all nodal values for temperature are initialized at ambient temperature, those nodes that have not yet been incorporated into the current dam's height will remain at ambient temperature. When the level of construction reaches these nodes, they will be at the temperature of the added concrete.

Note that the time associated with the heat of hydration is the current time less the time when the concrete was poured. This later time can be determined from the elevation of the point divided by the rate of construction. If this time is greater than the current time, the concrete has not yet been poured at that elevation. If it is less, then the difference between it and the current time is the time needed to determine H.

P3. **Consolidation of soil under an overburden.** When any structure is placed on a foundation of saturated soil, its weight is supported by an increase in the pore water pressure rather than the soil skeleton. This excess pore water pressure creates a flow away from the loaded area, which reduces the pressure and transfers the load to the soil skeleton. As the pore water pressure dissipates, the soil begins to consolidate, which gives rise to the time-dependent settlement of the structure. Geotechnical engineers must calculate the time-dependent drainage of the pore water in order to evaluate the settlement of the structure and possible foundation failure.

Consider the earth dam shown in the figure. Its foundation consists of two layers of saturated soil above impervious rock. It is reasonable to assume that shortly after construction of the dam, the excess pore water pressure at any point beneath the dam is directly proportional to the height of the dam above the point. Thus, the initial excess pressure at any point in the soil is equal to $\gamma_o h$, where γ_o is the unit weight of the overburden material and h is the height of the dam directly above the point.

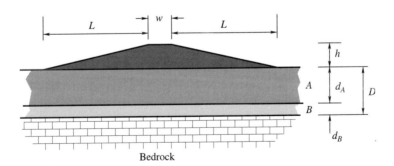

Bedrock

The governing equation for the time-dependent excess pore pressure is based on Darcy's law and was derived by Terzaghi. It can be simplified for this problem as

$$\frac{\partial}{\partial x}\left(c_v \frac{\partial u}{\partial x}\right) + \frac{\partial}{\partial y}\left(c_v \frac{\partial u}{\partial y}\right) = \frac{\partial u}{\partial t}$$

where c_v is the coefficient of consolidation and has dimensions $[L^2/T]$. It is a function of the soil's compressibility, void ratio, and permeability, as well as the specific weight of water. The equation can be considered nondimensional by using the following units:

Length	D
Time	$D^2/(c_v)_A$
Pressure	$\gamma_o h$

where $(c_v)_A$ is the value for soil A.
For the following parameters,

$(c_v)_B$	d_A	d_B	h	L	w
$20(c_v)_A$	$0.775D$	$0.225D$	$0.5D$	$4.5D$	D

determine the excess pore water pressure as a function of time. Assume no drainage into the impervious bedrock and sufficient drainage along the upper surface beneath the dam and at the toe and heel of the dam to maintain a zero value of excess pressure.

From the results of the preceding analysis, calculate how long it takes to reduce the maximum excess pressure to half the initial maximum excess pressure if $(c_v)_A = 7$ ft/year and $D = 20$ ft. This time should be large compared with the time it would take to construct the dam. In your opinion, would that be the case?

For your analysis, assume symmetry about the center of the dam and use a mesh that extends only to the toe of the dam. Here the initial excess pressure is zero and remains zero. This analysis will take several thousand increments of time, so you might consider conducting it in segments and using PHIstrt to initialize each new start.

It was research in solid mechanics that brought the finite element method to the forefront of engineering analysis, and this subject still accounts for much of the finite element research being conducted today. Although the focus of this chapter is on the analysis of linear elastic solids, in developing the concepts needed for these analyses the foundations will be laid for the development of other continuum mechanics problems, including Newtonian fluids. Among the new concepts introduced are: (1) the analysis of vector fields rather than scalar fields; (2) use of local, rotated coordinates; and (3) stress smoothing.

11.1 GENERAL DESCRIPTION OF PROBLEM

The basic problem of elasticity is to determine the displacement field within a solid due to prescribed forces and/or displacements acting on the solid. It is assumed that the solid is constrained to prevent pure rigid body motions. Associated with the displacements and forces are the secondary variables of stress and strain, related to each other by the constitutive equations for a linear elastic solid. The forces acting on the solid are of two types: body forces and surface tractions, the latter being the components of forces per unit area of surface. The following notation will be used throughout this chapter:

$$\{u\} = \begin{Bmatrix} u \\ v \\ w \end{Bmatrix} \qquad \{X\} = \begin{Bmatrix} X \\ Y \\ Z \end{Bmatrix} \qquad \{T\} = \begin{Bmatrix} T_x \\ T_y \\ T_z \end{Bmatrix}$$

Displacement vector   Body force vector   Surface traction vector

$$\{\sigma\} = \begin{Bmatrix} \sigma_{xx} \\ \sigma_{yy} \\ \sigma_{zz} \\ \sigma_{xy} \\ \sigma_{xz} \\ \sigma_{yx} \\ \sigma_{yz} \\ \sigma_{zx} \\ \sigma_{zy} \end{Bmatrix} \qquad \{\epsilon\} = \begin{Bmatrix} \epsilon_{xx} \\ \epsilon_{yy} \\ \epsilon_{zz} \\ \epsilon_{xy} \\ \epsilon_{xz} \\ \epsilon_{yx} \\ \epsilon_{yz} \\ \epsilon_{zx} \\ \epsilon_{zy} \end{Bmatrix}$$

Stress tensor   Strain tensor

where the stress and strain tensors are represented as vectors rather than using the more conventional matrix tensor notation. This vector notation is particularly convenient for the development of the finite element equations. Figure 11.1 illustrates the positive directions assumed for these quantities.

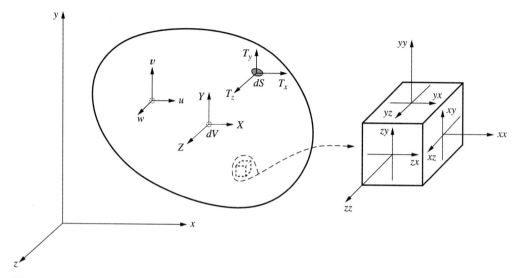

Figure 11.1. Displacements, body forces, surface tractions, and stress components.

11.2 CONSTITUTIVE EQUATION

The relationship between the stress and strain tensors for an isotropic linear elastic material is

$$
\begin{Bmatrix} \sigma_{xx} \\ \sigma_{yy} \\ \sigma_{zz} \\ \sigma_{xy} \\ \sigma_{xz} \\ \sigma_{yx} \\ \sigma_{yz} \\ \sigma_{zx} \\ \sigma_{zy} \end{Bmatrix} = \begin{bmatrix} A & B & B & 0 & 0 & 0 & 0 & 0 & 0 \\ B & A & B & 0 & 0 & 0 & 0 & 0 & 0 \\ B & B & A & 0 & 0 & 0 & 0 & 0 & 0 \\ 0 & 0 & 0 & C & 0 & 0 & 0 & 0 & 0 \\ 0 & 0 & 0 & 0 & C & 0 & 0 & 0 & 0 \\ 0 & 0 & 0 & 0 & 0 & C & 0 & 0 & 0 \\ 0 & 0 & 0 & 0 & 0 & 0 & C & 0 & 0 \\ 0 & 0 & 0 & 0 & 0 & 0 & 0 & C & 0 \\ 0 & 0 & 0 & 0 & 0 & 0 & 0 & 0 & C \end{bmatrix} \begin{Bmatrix} \epsilon_{xx} \\ \epsilon_{yy} \\ \epsilon_{zz} \\ \epsilon_{xy} \\ \epsilon_{xz} \\ \epsilon_{yx} \\ \epsilon_{yz} \\ \epsilon_{zx} \\ \epsilon_{zy} \end{Bmatrix} \tag{11.1}
$$

or, in abbreviated notation,

$$\{\sigma\} = [R]\{\epsilon\} \tag{11.2}$$

In Eq. 11.1

$$A = 2G + \lambda = \frac{E(1 - \nu)}{(1 + \nu)(1 - 2\nu)}$$

$$B = \lambda = \frac{\nu E}{(1 + \nu)(1 - 2\nu)} \tag{11.3}$$

$$C = 2G = \frac{E}{(1 + \nu)}$$

where
  ν is Poisson's ratio
  λ is Lamé's second constant
  E is Young's modulus of elasticity, or simply modulus of elasticity
  G is shear modulus of elasticity

11.3 CONDITION FOR EQUILIBRIUM

When the equations of equilibrium are applied to a differential element within an elastic body, the resulting equations represent the strong form for expressing equilibrium. When these equations are then placed in their weak form, the resulting equation is that of virtual work. For linear elastic solids acted on by conservative forces, virtual work can be shown to be the negative of the variation of the total potential energy of the system. Because virtual work, or the variation of the potential energy, is the most direct approach to the finite element equations, we begin with it. Symmetry of the stress tensor, as obtained from the conservation of angular momentum of a differential element, will be assumed.

The virtual work of all forces acting on the accumulation of particles that make up a solid can be shown to be

$$\delta W = -\int_V \lfloor \delta \epsilon \rfloor \{\sigma\} \, dV + \int_V \lfloor \delta u \rfloor \{X\} \, dV + \int_S \lfloor \delta u \rfloor \{T\} \, dS = 0 \tag{11.4}$$

which, for equilibrium, must equal zero as indicated. For elastic solids and conservative forces, this represents the negative of the variation of total potential energy,

$$\delta V = \int_V \lfloor \delta \epsilon \rfloor [R] \{\epsilon\} \, dV - \int_V \lfloor \delta u \rfloor \{X\} \, dV - \int_S \lfloor \delta u \rfloor \{T\} \, dS = 0 \tag{11.5}$$

where, for constant forces,

$$V = \frac{1}{2} \int_V \lfloor \epsilon \rfloor [R] \{\epsilon\} \, dV - \int_V \lfloor u \rfloor \{X\} \, dV - \int_S \lfloor u \rfloor \{T\} \, dS \tag{11.6}$$

Equation 11.5 is the starting point for the development of the finite element method for elastic solids.

11.4 STRAIN DISPLACEMENT RELATIONSHIPS

To formulate a finite element approximation to a problem in elasticity, it is be necessary to write the expression of virtual work in terms of displacements. This is accomplished through the use of the following strain displacement relationships:

$$
\begin{Bmatrix}
\epsilon_{xx} \\
\epsilon_{yy} \\
\epsilon_{zz} \\
\epsilon_{xy} \\
\epsilon_{xz} \\
\epsilon_{yx} \\
\epsilon_{yz} \\
\epsilon_{zx} \\
\epsilon_{zy}
\end{Bmatrix}
=
\begin{bmatrix}
\partial_x & 0 & 0 \\
0 & \partial_y & 0 \\
0 & 0 & \partial_z \\
\frac{1}{2}\partial_y & \frac{1}{2}\partial_x & 0 \\
\frac{1}{2}\partial_z & 0 & \frac{1}{2}\partial_x \\
\frac{1}{2}\partial_y & \frac{1}{2}\partial_x & 0 \\
0 & \frac{1}{2}\partial_z & \frac{1}{2}\partial_y \\
\frac{1}{2}\partial_z & 0 & \frac{1}{2}\partial_x \\
0 & \frac{1}{2}\partial_z & \frac{1}{2}\partial_y
\end{bmatrix}
\begin{Bmatrix}
u \\
v \\
w
\end{Bmatrix}
\tag{11.7}
$$

where

$$
\partial_x(.) = \frac{\partial(.)}{\partial x}, \text{ etc.}
$$

We will use the following abbreviated form for the above expression:

$$
\{\epsilon\} = [A]\{u\}
$$

11.5 PLANE STRAIN AND PLANE STRESS

Two-dimensional problems can be either plane strain problems or plane stress problems. Plane strain problems are those where the displacements normal to the plane of analysis are identically zero, i.e., $w \equiv 0$. Hence, the strain tensor is simply

$$
\{\epsilon\} =
\begin{Bmatrix}
\epsilon_{xx} \\
\epsilon_{yy} \\
\epsilon_{xy} \\
\epsilon_{yx}
\end{Bmatrix}
\tag{11.8}
$$

with all other components zero.

Plane stress problems are those where all components of stress in the third dimension are assumed zero. This occurs in the analysis of thin, plane plates loaded in the plane of the plate. In such cases the stress tensor reduces to

$$\{\epsilon\} = \begin{Bmatrix} \sigma_{xx} \\ \sigma_{yy} \\ \sigma_{xy} \\ \sigma_{yx} \end{Bmatrix} \tag{11.9}$$

with all other components zero.

For both plane strain and plane stress, the expression for virtual work reduces to

$$\int_V \lfloor \delta\epsilon_{xx} \quad \delta\epsilon_{yy} \quad \delta\epsilon_{xy} \quad \delta\epsilon_{yx} \rfloor \begin{Bmatrix} \sigma_{xx} \\ \sigma_{yy} \\ \sigma_{xy} \\ \sigma_{yx} \end{Bmatrix} dV =$$

$$\int_V \lfloor \delta u_x \quad \delta u_y \rfloor \begin{Bmatrix} X \\ Y \end{Bmatrix} dV + \int_V \lfloor \delta u_x \quad \delta u_y \rfloor \begin{Bmatrix} T_x \\ T_y \end{Bmatrix} dS \tag{11.10}$$

The constitutive relationship, Eq. 11.1, for plane strain and plane stress simplifies to

$$\begin{Bmatrix} \sigma_{xx} \\ \sigma_{yy} \\ \sigma_{xy} \\ \sigma_{yx} \end{Bmatrix} = \begin{bmatrix} 2G + \lambda & \lambda & 0 & 0 \\ \lambda & 2G + \lambda & 0 & 0 \\ 0 & 0 & 2G & 0 \\ 0 & 0 & 0 & 2G \end{bmatrix} \begin{Bmatrix} \epsilon_{xx} \\ \epsilon_{yy} \\ \epsilon_{xy} \\ \epsilon_{yx} \end{Bmatrix} \tag{11.11}$$

Plane strain

$$\begin{Bmatrix} \sigma_{xx} \\ \sigma_{yy} \\ \sigma_{xy} \\ \sigma_{yx} \end{Bmatrix} = \begin{bmatrix} 2G + \lambda - \lambda C & \lambda - \lambda C & 0 & 0 \\ \lambda - \lambda C & 2G + \lambda - \lambda C & 0 & 0 \\ 0 & 0 & 2G & 0 \\ 0 & 0 & 0 & 2G \end{bmatrix} \begin{Bmatrix} \epsilon_{xx} \\ \epsilon_{yy} \\ \epsilon_{xy} \\ \epsilon_{yx} \end{Bmatrix} \tag{11.12}$$

Plane stress

where

$$C = \frac{\lambda}{(2G + \lambda)}$$

The plane strain equation is straightforward. The plane stress terms are obtained by noting that when $\sigma_{zz} = 0.0$,

$$\lambda(\epsilon_{xx} + \epsilon_{yy}) + (2G + \lambda)\epsilon_{zz} = 0 \tag{11.13}$$

thus

$$\epsilon_{zz} = \frac{-\lambda}{(2G + \lambda)}(\epsilon_{xx} + \epsilon_{yy}) \tag{11.14}$$

Substitution of this into Eq. 11.1 leads directly to Eq. 11.12.

11.6 FINITE ELEMENT APPROXIMATIONS

The primary functions to be represented by our finite element approximations are the displacements. We consider here only two-dimensional problems; hence, we will associate with each node two displacements. For a single element,

$$u(x, y) = \lfloor N \rfloor \{U\}$$

$$v(x, y) = \lfloor N \rfloor \{V\} \tag{11.15}$$

where

$\{U\}$ is the array of nodal point displacements in the x direction
$\{V\}$ is the array of nodal point displacements in the y direction

This set of equations will also be written as

$$\left\{ \begin{array}{c} u(x, y) \\ v(x, y) \end{array} \right\} = \left[\begin{array}{cc} N & 0 \\ 0 & N \end{array} \right] \left\{ \begin{array}{c} U \\ V \end{array} \right\} \tag{11.16}$$

and abbreviated as

$$\{u\} = [N]\{U\} \tag{11.17}$$

The strains can now be approximated by

$$\{\epsilon\} = [A]\{u\} = [A][N]\{U\} \tag{11.18}$$

which gives us

$$\begin{Bmatrix} \epsilon_{xx} \\ \epsilon_{yy} \\ \epsilon_{xy} \\ \epsilon_{yx} \end{Bmatrix} = \begin{bmatrix} \left\lfloor \dfrac{\partial N}{\partial x} \right\rfloor & 0 \\ 0 & \left\lfloor \dfrac{\partial N}{\partial y} \right\rfloor \\ \dfrac{1}{2}\left\lfloor \dfrac{\partial N}{\partial y} \right\rfloor & \dfrac{1}{2}\left\lfloor \dfrac{\partial N}{\partial x} \right\rfloor \\ \dfrac{1}{2}\left\lfloor \dfrac{\partial N}{\partial y} \right\rfloor & \dfrac{1}{2}\left\lfloor \dfrac{\partial N}{\partial x} \right\rfloor \end{bmatrix} \begin{Bmatrix} \{U\} \\ \{V\} \end{Bmatrix} \tag{11.19}$$

The following abbreviations for the above expressions are used:

$$\begin{Bmatrix} \epsilon_{xx} \\ \epsilon_{yy} \\ \epsilon_{xy} \\ \epsilon_{yx} \end{Bmatrix} = \begin{bmatrix} N_x & 0 \\ 0 & N_y \\ \frac{1}{2}N_y & \frac{1}{2}N_x \\ \frac{1}{2}N_y & \frac{1}{2}N_x \end{bmatrix} \begin{Bmatrix} U \\ V \end{Bmatrix} \tag{11.20}$$

and

$$\{\epsilon\} = [N']\{U\} \tag{11.21}$$

11.7 THE FINITE ELEMENT EQUATIONS

We can now substitute the preceding approximations into the weak form of our governing equation, Eq. 11.5, to obtain the governing finite element equations. As usual, the substitution and integration are performed element by element, with each result assembled into the global matrices. For each element of V and each segment of S, we obtain

$$\int_{V_e} \lfloor \delta\epsilon \rfloor [R]\{\epsilon\}\, dV = \int_{V_e} \lfloor \delta U \rfloor [N']^T [R][N']\{U\}\, dV$$

$$= \lfloor \delta U \rfloor \left[\int_{V_e} [N']^T [R][N']\, dV \right]\{U\} = \lfloor \delta U \rfloor [K]_e\{U\} \tag{11.22}$$

$$\int_{V_e} \lfloor \delta u \rfloor \{X\}\, dV = \int_{V_e} \lfloor \delta U \rfloor [N]^T \{X\}\, dV = \lfloor \delta U \rfloor \left\{ \int_{V_e} [N]^T \{X\}\, dV \right\} = \lfloor \delta U \rfloor \{F\}_e \tag{11.23}$$

$$\int_{S_s} \lfloor \delta u \rfloor \{T\}\, dS = \int_{S_s} \lfloor \delta U \rfloor [N]^T \{T\}\, dS = \lfloor \delta U \rfloor \left\{ \int_{S_e} [N]^T \{T\}\, dS \right\} = \lfloor \delta U \rfloor \{F\}_s \tag{11.24}$$

After assembly into the global matrix equation, and recognition that this assembled equation must be true for arbitrary $\{\delta U\}$, we obtain

$$\sum_{1}^{\text{elements}} [K]_e\{U\} = \sum_{1}^{\text{elements}} \{F\}_e + \sum_{1}^{\text{segments}} \{F\}_s$$

or simply

$$[K]\{U\} = \{F\} \tag{11.25}$$

11.8 STRESS SMOOTHING

In elasticity it is not only the displacement field that is of interest to the analyst, but also the stress field. Because stress and strain are related to the gradient of displacement, they can be discontinuous across element boundaries. Even worse, there can be as many values of stress at a node as there are elements framing into the node. Therefore, how does one define nodal point stress and how does one plot stress? The answer is to apply what is known as *stress smoothing*, which determines a continuous stress field that is made close to the discontinuous field obtained from the displacement field.

Let σ be the discontinuous stress field (any component of the stress) found from the gradient of the displacement field, and let σ^* be an approximation to this field that is continuous across element boundaries (e.g., a finite element approximation to the desired stress field). Ideally, we would like

$$\sigma^* - \sigma = 0 \tag{11.26}$$

at all points in our mesh. In general, this will not be possible; hence, we seek to make this true in an average sense throughout our domain by demanding

$$\int_V W\{\sigma^* - \sigma\}\, dV = 0 \tag{11.27}$$

for as many independent weighting functions as we have nodal values for σ^*. This, of course, leads us to our standard finite element equations whereby we interpret the W functions as arbitrary variations of our nodal stresses, that is,

$$\int_V \delta\sigma^*\{\sigma^* - \sigma\}\, dV = 0 \tag{11.28}$$

Thus, by substituting our finite element approximations for σ^*, we obtain

$$\int_{V_e} \lfloor\delta\sigma^*\rfloor\{N\}\lfloor N\rfloor\{\sigma^*\}\, dV = \int_{V_e} \lfloor\delta\sigma^*\rfloor\{N\}\{\sigma\}\, dV \tag{11.29}$$

or

$$[M]\{\sigma^*\} = \{F_\sigma\} \tag{11.30}$$

where

$$[M] = \sum \int_{V_e} \{N\} \lfloor N \rfloor \, dV \tag{11.31}$$

$$\{F_\sigma\} = \sum \int_{V_e} \{N\} \sigma \, dV \tag{11.32}$$

The matrix $[M]$ is identical to the capacitance matrix encountered in transient analysis. We discovered at that time that it can be diagonalized to give accurate results without the additional computer time required to factor a matrix. The same is true for stress smoothing. Therefore, the usual procedure is to diagonalize this matrix by one of the methods previously discussed and solve for nodal point stresses, component by component, in an explicit manner. (Note that the terms in $[M]$ are the same regardless of which component of stress is being smoothed.)

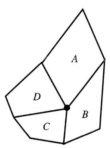

Figure 11.2. A node for stress smoothing.

It can be shown that our previous method for lumping matrices gives weights proportional to the size of the elements at a given node. This may not be the best approach for smoothing stresses. For example, in Fig. 11.2, element A has the largest area, but its centroid is farthest from the node and it also represents the smallest fraction of the total angle surrounding the node. Therefore, we might consider a second method that weights the contribution from a given element using the angle formed by its two sides at the node. If that were used, then element B would contribute the most to determining the stress at our particular node. Finally, a third method could use the inverse of the distance to the centroid (or at each quadrature point) of each of the elements. This gives the largest weight to the element whose centroid is closest to the node. If this were employed, the largest weight would go to element B or C; they would be about even. Any of these methods can be used to give smooth stress components that converge to the correct values as the displacements converge to the true displacements. The reader is encouraged to experiment with all of these methods.

11.9 PROGRAMMING PRELIMINARIES

Because the code for elastic solids has as its primary variable a vector field, as opposed to our previous scalar field problems, there are two new programming details: (1) storage of the matrix equations and (2) specification of boundary conditions.

11.9.1 Stiffness Array and Storage. As with our other programs, the integration of the element stiffness matrices, as well as the forcing functions, will be done using Gaussian quadrature. However, because there are two variables per node, one for each coordinate direction, some care must be taken in terms of organization. We will find it convenient to divide the element stiffness matrix into four submatrices:

$$\lfloor \delta U \rfloor [K]_e \{U\} = \lfloor \delta U \quad \delta V \rfloor \begin{bmatrix} K_{xx} & K_{xy} \\ K_{yx} & K_{yy} \end{bmatrix}_e \begin{Bmatrix} U \\ V \end{Bmatrix} \tag{11.33}$$

Also, because the matrix is symmetric,

$$[K_{yx}] = [K_{xy}]^T \tag{11.34}$$

and only $[K_{xy}]$ need be calculated. Expansion of Eq. 11.22 gives

$$[K_{xx}] = \int_{V_e} \left[\{N_x\} R_{11} \lfloor N_x \rfloor + \{N_y\} \frac{R_{33}}{2} \lfloor N_y \rfloor \right] dV \tag{11.35}$$

$$[K_{xy}] = \int_{V_e} \left[\{N_x\} R_{12} \lfloor N_y \rfloor + \{N_y\} \frac{R_{33}}{2} \lfloor N_x \rfloor \right] dV \tag{11.36}$$

$$[K_{yy}] = \int_{V_e} \left[\{N_y\} R_{22} \lfloor N_y \rfloor + \{N_x\} \frac{R_{33}}{2} \lfloor N_x \rfloor \right] dV \tag{11.37}$$

Here the R values shown are those appropriate for either plane strain or plane stress as given by Eqs. 11.11 and 11.12. Also, in writing the above expressions, the equalities $R_{12} = R_{21}$ and $R_{33} = R_{44}$ were assumed. If n represents the number of nodes per element, then the size of each of the above matrices is $n \times n$, and the total element stiffness matrix shown in Eq. 11.33 is $2n \times 2n$.

Now that we have expressions for each term in the element stiffness matrix, we turn our attention to the assembly of these terms into the global stiffness matrix. To minimize the bandwidth, it will be necessary to organize the global stiffness matrix differently than the element stiffness matrix. Rather than grouping all the x components together and all the y components together, the x and y components for each node will be grouped together. Hence, the displacement vector and corresponding right-hand side will be written as

$$
\left\{
\begin{array}{c}
U_1 \\
\hline
V_1 \\
\hline
U_2 \\
\hline
V_2 \\
\hline
U_3 \\
\hline
V_3 \\
\hline
U_4 \\
\hline
V_4 \\
\hline
\vdots \\
\hline
U_N \\
\hline
V_N
\end{array}
\right\}
\qquad
\left\{
\begin{array}{c}
FX_1 \\
\hline
FY_1 \\
\hline
FX_2 \\
\hline
FY_2 \\
\hline
FX_3 \\
\hline
FY_3 \\
\hline
FX_4 \\
\hline
FY_4 \\
\hline
\vdots \\
\hline
FX_N \\
\hline
FY_N
\end{array}
\right\}
\tag{11.38}
$$

With this arrangement, an x component of a vector that is associated with node n will occupy position $(2n - 1)$ and the y component will occupy position $(2n)$. This arrangement requires that terms in the element stiffness matrix be stored in the noncompact global stiffness matrix using the following algorithm:

Term	Row	Column
$K_{xx}(I, J)$	$2I - 1$	$2J - 1$
$K_{yy}(I, J)$	$2I$	$2J$
$K_{xy}(I, J)$	$2I - 1$	$2J$
$K_{yx}(I, J)$	$2I$	$2J - 1$

Figure 11.3 illustrates this arrangement for the four terms associated with row m and column n.

The physical interpretation that can be given to each the four terms is as follows:

$K_{xx}(m, n)$ x component of force at node m due to a unit displacement in the x direction at node n

$K_{xy}(m, n)$ x component of force at node m due to a unit displacement in the y direction at node n

$K_{yx}(m, n)$ y component of force at node m due to a unit displacement in the x direction at node n

$K_{yy}(m, n)$ y component of force at node m due to a unit displacement in the y direction at node n

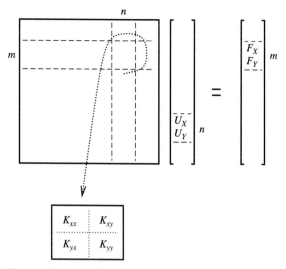

Figure 11.3. Stiffness matrix storage arrangement.

11.9.2 Boundary Conditions and Local, Rotated Coordinates. In the previous finite element formulations we noted that for each equation either the forcing function, or the primary variable, or the relationship between them must be specified for each equation. Other conditions could also have applied, such as some relationship between two or more nodal variables. This latter condition, which was possible in many of our previous applications, now becomes probable.

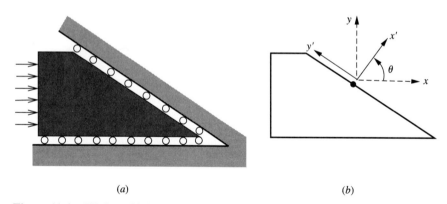

Figure 11.4. Wedge with local coordinates.

Take, for example, the problem illustrated in Fig. 11.4a. Along the side in contact with the sloping wall, we know the displacement normal to the wall and the traction parallel to the wall (considered frictionless). However, we know neither the displacement in the x direction nor that in the y direction, but rather their ratios. We can express these known ratios as additional equations, but this is not convenient. A much more practical solution is to assign, to each node, a local coordinate system associated with the direction of known displacement (see Fig. 11.4*b*). If θ is the counterclockwise rotation of the x axis into the x' axis, then we can represent the u', v' displacements in terms of the u, v displacements as

$$\left\{ \begin{array}{c} u \\ v \end{array} \right\} = \left[\begin{array}{cc} \cos(\theta) & -\sin(\theta) \\ \sin(\theta) & \cos(\theta) \end{array} \right] \left\{ \begin{array}{c} u' \\ v' \end{array} \right\} \tag{11.39}$$

We now define a global transformation matrix

$$[T] = \left[\begin{array}{ccccccccc} 1 & & & & & & & & \\ & 1 & & & & & & & \\ & & 1 & & & & & & \\ & & & 1 & & & & & \\ & & & & 1 & & & & \\ & & & & & C & -S & & \\ & & & & & S & C & & \\ & & & & & & & 1 & \\ & & & & & & & & 1 \\ & & & & & & & & & 1 \end{array} \right] \tag{11.40}$$

where C and S represent the sine and cosine of the angle associated with the nodal point being considered, and write our global finite element equation as

$$[K][T]\{U'\} = [T]\{F'\} \tag{11.41}$$

Note that $\{U'\}$ and $\{F'\}$ are the components of displacement and force with respect to whichever coordinate system is associated with the node. For most nodes, this will be the global coordinate system (i.e., the angle θ will be zero).

Because the transformation matrix $[T]$ is orthogonal—that is, its inverse is equal to its transpose—we can write

$$[T]^T[K][T]\{U'\} = \{F'\} \tag{11.42}$$

$$[K']\{U'\} = \{F'\}$$

where $[K'] = [T]^T[K][T]$. It should be understood that the $[T]$ matrix can contain rotations for more than one node; in fact, it should contain the rotations associated with all nodes where local coordinates have been assigned. In a computer code, the transformations are performed at the element level rather than at the global level; hence, when the element stiffness matrix and right-hand side are assembled into the global matrix, the terms are already given with respect to locally defined coordinate systems. Now, because the displacements and forces are expressed with respect to the local coordinates used at each node, their components can be prescribed as we have always done.

Finally, in order to make use of this procedure, it will be necessary to specify for each node the angle θ. This specification will be made in the user's INCLUDE file, INITIAL.m. If this angle is zero, the program can skip the transformation; otherwise, it will perform the transformation. Output can be given in either local or global coordinates or both.

Two final comments on the use of local coordinates: (1) They do not have to be oriented normal and tangential to the boundary. Although that will be the most common use for them, the user is free to use them in any capacity that is convenient. (2) They can be used for nodes other than those on the boundary. Again, boundary specifications will be the most common use for them, but there is nothing to prohibit their use at an interior node.

11.10 PROGRAM elastic.m

Program elastic.m is an executive program that calls two other finite element programs: strain.m for calculating nodal point displacements, and stress.m for calculating nodal point stresses based on the displacements obtained from strain.m. Thus, the following programs and data files are needed in your working directory to run an elasticity analysis:

Primary Codes	Input Data	User's INCLUDEs	Supplied Functions
elastic.m	MESHo	INITIAL.m	SFquad.m
strain.m	NODES	COEF.m	sGAUSS.m
stress.m	NP		
	NWLD		

11.10.1 Flow Chart: elastic.m. Program elastic.m is the "executive" program for the complete set.

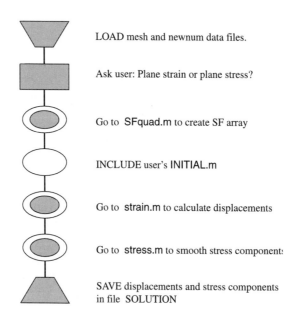

LOAD mesh and newnum data files.

Ask user: Plane strain or plane stress?

Go to SFquad.m to create SF array

INCLUDE user's INITIAL.m

Go to strain.m to calculate displacements

Go to stress.m to smooth stress components

SAVE displacements and stress components in file SOLUTION

11.10.2 Code: elastic.m

```
%-------------------
% elastic.m
%-------------------
 clear

 load MESHo   -ASCII
 load NODES   -ASCII
 load NP      -ASCII
 load NWLD    -ASCII

 NUMNP = MESHo(1);
 NUMEL = MESHo(2);
 NNPE  = MESHo(3);
 for I=1:NUMNP;
   XORD(I)  =NODES(I,1);
   YORD(I)  =NODES(I,2);
   NPcode(I)=NODES(I,3);
 end

 if NNPE      == 3
    NSPE=3;    NNPS=2;
 elseif NNPE == 6
    NSPE=3;    NNPS=3;
 elseif NNPE == 4
    NSPE=4;    NNPS=2;
 elseif NNPE == 8
    NSPE=4;    NNPS=3;
 end

 IB    = 2*NWLD(NUMNP+1);
 NUMEQ = 2*NUMNP;
 for I = 1:NUMNP
    NPBC(I)    = 0;
    XBC(I)     = 0.0;
    YBC(I)     = 0.0;
    TX(I)      = 0.0;
    TY(I)      = 0.0;
    THETA(I)   = 0.0;
 end
 for I = 1:NUMEQ
    for J = 1:IB
       SK(I,J)=0;
    end
 end
```

This is the executive program that loads external data, calls the programs for displacement and stress, and saves results in file RESULTS.

Load data from mesh.m and newnum.m. All four files must be in working directory.

File	Contents
MESHo	NUMNP, NUMEL, NNPE
NODES	XORD, YORD, NPBC
NP	NP array
NWLD	New nodal numbers
	Last entry is IB

Variable	Definition
NUMNP	Number of nodal points
NUMEL	Number of elements
NNPE	Number of nodes per element
XORD	Node's x coordinates
YORD	Node's y coordinates
NPcode	Nodal point code
NSPE	Number of sides per element
NNPS	Number of nodes per side

Convert IB defined by newnum.m to IB for a symmetric analysis with two variables per node.

Initialize and set default values.

Variable	Definition
NPBC	Nodal point boundary condition code
XBC, YBC	Components of known boundary condition either force or displacement
TX, TY	Components of surface traction
THETA	Angle for rotated local coordinate
SK	Global stiffnes matrix

```
%-------------------------------------
%  Ask user for type of analysis desired
%-------------------------------------
 disp(' ')
 disp(' ENTER:')
 disp(' --------------------')
 disp(' 1 for plane strain   ')
 disp(' 2 for plane stress   ')
 disp(' --------------------')
 IPROG = input(' < ');


 [SF,WT,NUMQPT,NPSIDE] = SFquad(NNPE);



%-----------------------------
% User-written initialization
%-----------------------------
   INITIAL



%-------------------
% SOLVE PROBLEM
%-------------------
 strain
 stress



%----------------------
% SAVE SOLUTION
%----------------------
 for I=1:NUMNP
   Ix = 2*I-1;
   Iy = 2*I;
   SOLUTION(I,1) = U(Ix);
   SOLUTION(I,2) = U(Iy);
   SOLUTION(I,3) = SIGXX(I);
   SOLUTION(I,4) = SIGXY(I);
   SOLUTION(I,5) = SIGYY(I);
   SOLUTION(I,6) = SGEFF(I);
 end


 save SOLUTION SOLUTION -ASCII
```

Ask user if analysis is plane strain or plane stress.

Go to **SFquad.m** to create SF array (see Appendix E).

Go to user's **INITIAL.m** to set boundary conditions.

Go to **strain.m** to calculate nodal displacements.
Go to **stress.m** to calculate nodal stresses.

Place all results in **SOLUTION.**

Save **SOLUTION** for plotting, etc.

```
NOTE = [
'------------------------------------';
' All nodal point values are saved in  ';
'           file SOLUTION              ';
'    located in your home directory.   ';
'------------------------------------';
'          Contents of SOLUTION        ';
'                                      ';
'    UX UY SIGXX SIGXY SIGYY SGEFF     ';
'where:                                ';
'   UX, UY       = x and y displacements ';
'   SIGXX, etc.  = Stress components   ';
'   SGEFF        = Effective stress    ';
'------------------------------------';
'to retrieve, for example, SIGXX values,';
'     load SOLUTION                    ';
'     SIGXX = SOLUTION(1:NUMNP,3)      ';
'     save SIGXX SIGXX -ASCII          ';
'------------------------------------';
];

disp(NOTE)
```

Describe to user the file SOLUTION and how to use it.

11.10.3 Flow Chart: strain.m. Program strain.m calculates the nodal point displacements of the elastic solid.

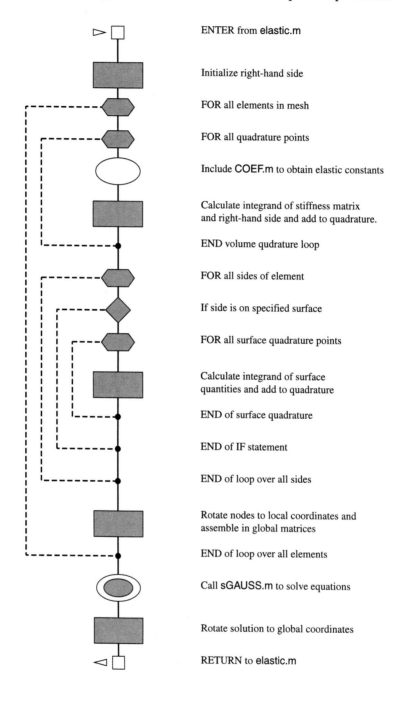

ENTER from elastic.m

Initialize right-hand side

FOR all elements in mesh

FOR all quadrature points

Include COEF.m to obtain elastic constants

Calculate integrand of stiffness matrix
and right-hand side and add to quadrature.

END volume qudrature loop

FOR all sides of element

If side is on specified surface

FOR all surface quadrature points

Calculate integrand of surface
quantities and add to quadrature

END of surface quadrature

END of IF statement

END of loop over all sides

Rotate nodes to local coordinates and
assemble in global matrices

END of loop over all elements

Call sGAUSS.m to solve equations

Rotate solution to global coordinates

RETURN to elastic.m

11.10.4 Code: strain.m

```
%==================
% ROUTINE strain.m
%==================
    NNPE2=2*NNPE;
    NUMEQ=2*NUMNP;


    for I=1:NUMNP;
        NI=NWLD(I);
        NIX=2*NI-1;
        NIY=NIX+1;
        F(NIX)=XBC(I);
        F(NIY)=YBC(I);
    end

% -------------------------------
% FORMATION OF ELEMENT S MATRICES
% -------------------------------
    for I=1:NUMEL

        for J=1:NNPE2
            FE(J)=0.0;
            for K=1:NNPE2
                SE(J,K)=0.0;
            end
        end

        for J=1:NNPE
            NPJ=NWLD(NP(I,J));
            NPE(J)=NPJ*2-1;
            J1=J+NNPE;
            NPE(J1)=NPJ*2;
        end

% -------------------------
% BEGIN VOLUME QUADRATURE
% -------------------------
        JEND=NUMQPT(1);
        for J=1:JEND
            XJ        =0.0;
            YJ        =0.0;
            RJAC(1,1)=0.0;
            RJAC(1,2)=0.0;
            RJAC(2,1)=0.0;
            RJAC(2,2)=0.0;
```

Routine strain.m.
Called from elastic.m to calculate nodal point displacements

Variable	Definition
NNPE2	2(number of nodes per element)
NUMEQ	Number of equations
NI	New node number for node *I*
NIX	Equation numbers associated with
NIY	*x* and *y* components, node *I*
XBC	Specified nodal force components
YBC	in local coordinates. *Note:*
	These were specified by user in
	INITIAL.m.

Begin creating global matrices and vectors element by element.

Initialize element stiffness and force matrices.

NPE = equation number of *x* component for each node of element using new nodal numbers

Begin Gaussian quadrature.

NUMQPT(1) = number of Gauss points for element quadrature

Initialize quadrature point coordinates.
Initialize Jacobian matrix.

$$\left[J\frac{(x, y)}{(u, v)} \right] = 0$$

```
for K=1:NNPE
  NPK=NP(I,K);
  XJ = XJ + SF(1,K,J)*XORD(NPK);
  YJ = YJ + SF(1,K,J)*YORD(NPK);
  RJAC(1,1)=RJAC(1,1)+SF(2,K,J)*XORD(NPK);
  RJAC(1,2)=RJAC(1,2)+SF(3,K,J)*XORD(NPK);
  RJAC(2,1)=RJAC(2,1)+SF(2,K,J)*YORD(NPK);
  RJAC(2,2)=RJAC(2,2)+SF(3,K,J)*YORD(NPK);
end
DETJ=RJAC(1,1)*RJAC(2,2)- ...
      RJAC(2,1)*RJAC(1,2);

if DETJ <= 0.0
   error
end

RJACI(1,1)=+RJAC(2,2)/DETJ;
RJACI(1,2)=-RJAC(1,2)/DETJ;
RJACI(2,1)=-RJAC(2,1)/DETJ;
RJACI(2,2)=+RJAC(1,1)/DETJ;

for K=1:NNPE
   DNDX(K)=RJACI(1,1)*SF(2,K,J)+ ...
           RJACI(2,1)*SF(3,K,J);
   DNDY(K)=RJACI(1,2)*SF(2,K,J)+ ...
           RJACI(2,2)*SF(3,K,J);
end
% -------------------------------
% Include user-written COEF.m code
% -------------------------------
 COEF

if IPROG == 1
   % Plane strain
   R(1,1)=2.0*G+LAMBDA;
   R(2,2)=R(1,1);
   R(3,3)=G;
   R(1,2)=LAMBDA;
   R(2,1)=LAMBDA;
elseif IPROG == 2
   % Plane stress
   R(1,1)=4.0*G*(G+LAMBDA)/(2.0*G+LAMBDA);
   R(1,2)=2.0*G*LAMBDA/(2.0*G+LAMBDA);
   R(2,1)=R(1,2);
   R(2,2)=R(1,1);
   R(3,3)=G;
end
```

Calculate quadrature point values:

$$XJ = \lfloor N \rfloor \{X\} \qquad YJ = \lfloor N \rfloor \{Y\}$$

$$RJAC(1,1) = \partial x/\partial u = \lfloor N_u \rfloor \{X\}$$
$$RJAC(1,2) = \partial x/\partial v = \lfloor N_v \rfloor \{X\}$$
$$RJAC(2,1) = \partial y/\partial u = \lfloor N_u \rfloor \{Y\}$$
$$RJAC(2,2) = \partial y/\partial v = \lfloor N_v \rfloor \{Y\}$$

$$DETJ = \|RJAC\|$$

If determinant is less than or equal to zero, there is an error in element mapping.

$$[RJACI] = [RJAC]^{-1}$$

$$\left[J \frac{(u, v)}{(x, y)} \right] = \left[J \frac{(x, y)}{(u, v)} \right]^{-1}$$

$$\begin{bmatrix} \left| \dfrac{\partial N}{\partial x} \right| \\ \left| \dfrac{\partial N}{\partial y} \right| \end{bmatrix} = J \frac{(u, v)}{(x, y)} \begin{bmatrix} \left| \dfrac{\partial N}{\partial u} \right| \\ \left| \dfrac{\partial N}{\partial v} \right| \end{bmatrix}$$

Include user-written COEF.m to obtain

GAMX = body force in x direction
GAMY = body force in y direction
G = shear modulus of elasticity
LAMBDA = Lamé's (second) constant

Fill in R matrix according to type of analysis desired, i.e., plane strain or plane stress.

```
%-------------------------------------------
%  CALCULATE QUADRATURE CONTRIBUTION
%  TO SE MATRIX
%-------------------------------------------
    for K=1:NNPE
        SFK=SF(1,K,J);
        K1=K;
        K2=K+NNPE;
        for L=1:NNPE
            L1=L;
            L2=L+NNPE;

            SE(K1,L1)=SE(K1,L1)+WT(1,J)...
                *(DNDX(K)*R(1,1)*DNDX(L)...
                + DNDY(K)*R(3,3)*DNDY(L))*DETJ;


            SE(K1,L2)=SE(K1,L2)+WT(1,J)...
                *(DNDX(K)*R(1,2)*DNDY(L)...
                + DNDY(K)*R(3,3)*DNDX(L))*DETJ;


            SE(K2,L2)=SE(K2,L2)+WT(1,J)...
                *(DNDY(K)*R(2,2)*DNDY(L)...
                + DNDX(K)*R(3,3)*DNDX(L))*DETJ;


            SE(L2,K1)=SE(K1,L2);
        end


        F(K1)=F(K1)+SFK*GAMX*DETJ;
        F(K2)=F(K2)+SFK*GAMY*DETJ;
    end

 end
%-------- end of volume quadrature
```

For each node of element,

SFK = $N(J)$ at current quadrature point
K1 = row number in $[S]_e$ for x values
K2 = row number in $[S]_e$ for y values

L1 = column number in $[S]_e$ for x values
L2 = column number in $[S]_e$ for y values

$[SE(K1, L1)] = [K_{xx}(K, L)]$

$$= \int_{V_e} \left[\{N_x\}R_{11}\lfloor N_x \rfloor + \{N_y\}R_{33}\lfloor N_y \rfloor \right] dV$$

$[SE(K1, L2)] = [K_{xy}(K, L)]$

$$= \int_{V_e} \left[\{N_x\}R_{12}\lfloor N_y \rfloor + \{N_y\}R_{33}\lfloor N_x \rfloor \right] dV$$

$[SE(K2, L2)] = [K_{yy}(K, L)]$

$$= \int_{V_e} \left[\{N_y\}R_{22}\lfloor N_y \rfloor + \{N_x\}R_{33}\lfloor N_x \rfloor \right] dV$$

$$[K_{yx}] = [K_{xy}^T]$$

$$FE(K1) = \int_{V_e} \{N\}X \, dV$$

$$FE(K2) = \int_{V_e} \{N\}Y \, dV$$

End of volume quadrature of current element

```
%-----------------------------
% BEGIN SURFACE QUADRATURE ON
% EACH SIDE OF ELEMENT I
%-----------------------------

  for J=1:NSPE
% ------------------------------
% CHECK IF QUADRATURE IS NECESSARY
% ------------------------------
    CHKTX=1.0;
    CHKTY=1.0;
    for K=1:NNPS
       J1=NP(I,NPSIDE(J,K));
       CHKTX=CHKTX*TX(J1);
       CHKTY=CHKTY*TY(J1);
    end

    if CHKTX ~= 0 |  CHKTY ~= 0
%      -----------------------------------
%      BEGIN SURFACE QUADRATURE ON SIDE J
%      -----------------------------------
%      > Note: Input data is rotated to  <
%      >       global coordinates for    <
%      >       integration.              <
%      -----------------------------------


       KEND=NUMQPT(2);
       for K=1:KEND
%         -----------------------
%         DETERMINE PARAMETERS AT
%         QUADRATURE POINT
%         -----------------------
          DXDXI=0.0;
          DYDXI=0.0;
          TXK=0.0;
          TYK=0.0;
          UXK=0.0;
          UYK=0.0;
```

NSPE = number of sides per element

On any given side of the element, at least one of the components of surface traction must be nonzero at all nodes if there is a surface integral to be evaluated.

CHKTX and CHKTY = check values for TX and TY

If CHKTX and CHKTY are both zero, then there is no need for surface quadrature.

Note: Components of nodal point surface tractions are entered by user in local coordinates. However, during an integration, all components must be given with respect to the same coordinate system. Hence, these components will be rotated to the global system before integration.

Initialize all variables to be evaluated to zero.

Note: XI = ξ = coordinate specifying position along boundary in the *x-y* plane

```
    for L=1:NNPS
      L1=NPSIDE(J,L);
      NPL=NP(I,L1);
      TH=THETA(NPL);
      C=cos(TH);
      S=sin(TH);


      DXDXI=DXDXI+SF(5,L,K)*XORD(NPL);
      DYDXI=DYDXI+SF(5,L,K)*YORD(NPL);


      TXK=TXK+SF(4,L,K)*(C*TX(NPL)...
                        -S*TY(NPL));
      TYK=TYK+SF(4,L,K)*(S*TX(NPL)...
                        +C*TY(NPL));
    end
```

NNPS = number of nodes per side
NPSIDE(J,L) = address in NP array of the *L*th node on side *J*
THETA(NPL) = angle local coordinate makes with global coordinates

$$\begin{Bmatrix} \partial x / \partial \xi \\ \partial y / \partial \xi \end{Bmatrix} = \begin{bmatrix} \partial N_s / \partial \xi \\ \partial N_s / \partial \xi \end{bmatrix} \begin{Bmatrix} \text{XORD} \\ \text{YORD} \end{Bmatrix}$$

$$\begin{Bmatrix} T_x \\ T_y \end{Bmatrix} = \begin{bmatrix} C & -S \\ S & C \end{bmatrix} \begin{bmatrix} N_s \\ N_s \end{bmatrix} \begin{Bmatrix} \text{TX} \\ \text{TY} \end{Bmatrix}$$

where $\partial N_s / \partial \xi$ and N_s are row vectors associated with one-dimensional shape functions, and XORD, YORD, TX, and TY are column vectors of nodal values. See Chapter 7 for complete details.

T_x, T_y, u, and v are scalar components of traction and force/displacement in global coordinates at the current quadrature point.

```
%       -------------------------
%       Quadrature contribution of
%       surface traction
%       -------------------------
      DETJS=sqrt(DXDXI^2+DYDXI^2);
      for L=1:NNPS
        COMM=WT(2,K)*SF(4,L,K)*DETJS;
        L1=NPSIDE(J,L);
        L2=L1+NNPE;
        FE(L1)=FE(L1)+ COMM*TXK;
        FE(L2)=FE(L2)+ COMM*TYK;
      end


    end
%    ---------- end of surface quadrature
  end
%  ---------- end of IF statement
 end
%---------- end of loop over sides
```

$$\frac{ds}{d\xi} = \left[\left(\frac{dx}{d\xi} \right)^2 + \left(\frac{dy}{d\xi} \right)^2 \right]^{1/2}$$

$$\begin{Bmatrix} F_x \\ F_y \end{Bmatrix} = \int_S \begin{bmatrix} N & 0 \\ 0 & N \end{bmatrix} \begin{Bmatrix} T_x \\ T_y \end{Bmatrix} \frac{ds}{d\xi} d\xi$$

```
%    -----------------------------
%    ROTATE SE and FE TO X'-Y' AXES
%        (Local Coordinates)
%    -----------------------------
     for J=1:NNPE
       TH=THETA(NP(I,J));
       if TH ~= 0
         C=cos(TH);
         S=sin(TH);


         J1=J;
         J2=J+NNPE;
         FEX =  C*FE(J1) + S*FE(J2);
         FEY = -S*FE(J1) + C*FE(J2);
         FE(J1) = FEX;
         FE(J2) = FEY;

         for K=1:NNPE;
           K1=K;
           K2=K+NNPE;
           XX=+C*SE(J1,K1)+S*SE(J2,K1);
           XY=+C*SE(J1,K2)+S*SE(J2,K2);
           YX=-S*SE(J1,K1)+C*SE(J2,K1);
           YY=-S*SE(J1,K2)+C*SE(J2,K2);
           SE(J1,K1)=XX;
           SE(J1,K2)=XY;
           SE(J2,K1)=YX;
           SE(J2,K2)=YY;
           XX=+SE(K1,J1)*C+SE(K1,J2)*S;
           XY=-SE(K1,J1)*S+SE(K1,J2)*C;
           YX=+SE(K2,J1)*C+SE(K2,J2)*S;
           YY=-SE(K2,J1)*S+SE(K2,J2)*C;
           SE(K1,J1)=XX;
           SE(K1,J2)=XY;
           SE(K2,J1)=YX;
           SE(K2,J2)=YY;
         end

       end
%        -------- end of IF statement
     end
%    ----- end of loop over NNPE
```

At this point, the element stiffness matrix and the element force vector are given with respect to global coordinates. Before they are assembled into the global stiffness matrix and force vector, their components must be rotated to local coordinates where specified.

NNPE = number of nodes per element

THETA = counterclockwise rotation of global coordinate to local coordinate

$$\{F'\} = [T]^{T}\{F\}$$

$$\begin{Bmatrix} F'_x \\ F'_y \end{Bmatrix} = \begin{bmatrix} C & S \\ -S & C \end{bmatrix} \begin{Bmatrix} F_x \\ F_y \end{Bmatrix}$$

$$[K'] = [T]^{T}[K][T]$$

$$\begin{bmatrix} S'_{xx} & S'_{xy} \\ S'_{yx} & S'_{yy} \end{bmatrix} = \begin{bmatrix} C & -S \\ S & C \end{bmatrix} \begin{bmatrix} S'_{xx} & S'_{xy} \\ S'_{yx} & S'_{yy} \end{bmatrix} \begin{bmatrix} C & S \\ -S & C \end{bmatrix}$$

```
%       -----------------------------------
%       PLACE SE and FE IN GLOBAL SK MATRIX
%       -----------------------------------
        for J=1:NNPE2
          NPJ=NPE(J);
          F(NPJ)=F(NPJ) + FE(J);
          for K=1:NNPE2
            NPK=NPE(K);
            if NPK >= NPJ
               JK=(NPK-NPJ)+1;
               SK(NPJ,JK)=SK(NPJ,JK)+SE(J,K);
            end
          end
        end

    end
%   --------   End of loop over elements.
%              All elements are now assembled.




%       ----------------------------
%       SPECIFY BOUNDARY CONDITIONS
%       ----------------------------
        for I=1:NUMNP
          NBCI=NPBC(I);
          if NBCI == 3 | NBCI == 1
             NPX=2*NWLD(I)-1;
             SK(NPX,1)=SK(NPX,1)*1.0E+12;
             F(NPX) =SK(NPX,1)*XBC(I);
          end

          if NBCI == 3 | NBCI == 2
             NPY=2*NWLD(I);
             SK(NPY,1)=SK(NPY,1)*1.0E+12;
             F(NPY) =SK(NPY,1)*YBC(I);
          end
        end
```

The global stiffness matrix is stored as a banded symmetric matrix.

NNPE2 = 2(number of nodes per element) = number of equations per element

NPE(J) = global equation number for node J

Store only diagonal and upper triangular terms, that is, where the column number, NPK, is equal to or larger than the row number, NPJ.

JK = column number for banded symmetric storage.

NPBC = 1 or 3: x component of displacement is specified. XBC = specified displacement.

NPBC = 2 or 3: y component of displacement is specified. YBC = specified displacement.

Decouple diagonal term from the off-diagonal terms by increasing its value several orders of magnitude.

Place known displacement, multiplied by the large diagonal term, on right-hand side.

Note: XBC and YBC are given in terms of the local coordinate system specified.

```
% -----------------------------
% SOLVE MATRIX EQUATION
% -----------------------------
   U =  sGAUSS(SK,F,NUMEQ,IB);


% -----------------------------
% PLACE SOLUTION IN ORDER OF
%      ORIGINAL NODE NUMBERS
%      AND GLOBAL COORDINATES
% -----------------------------
   for I=1:NUMEQ
     F(I)=U(I);
   end
   for I=1:NUMNP
     IY =2*I;
     IX =IY-1;
     TH =THETA(I);
     C  =cos(TH);
     S  =sin(TH);
     NI =NWLD(I);
     NIY=2*NI;
     NIX=NIY-1;
     U(IX)=F(NIX)*C-F(NIY)*S;
     U(IY)=F(NIX)*S+F(NIY)*C;
   end

% -------------------
% Return to elastic.m
% -------------------
```

sGAUSS.m is the Gauss elimination code for symmetric, banded matrices.

U is the vector of nodal point displacements. At this time, it is in order of the new numbering system and components in local coordinates.

Transfer U values to a new storage area to prepare for placing results in order of original nodal numbering, rotated to global coordinates.

NWLD = new nodal number in terms of old nodal number

U now contains displacement components (global coordinates) in order of original (from mesh.m) nodal numbers.

Nodal displacement values are now known. Return to elastic.m.

11.10.5 Flow Chart: stress.m. Program stress.m calculates nodal point stress components.

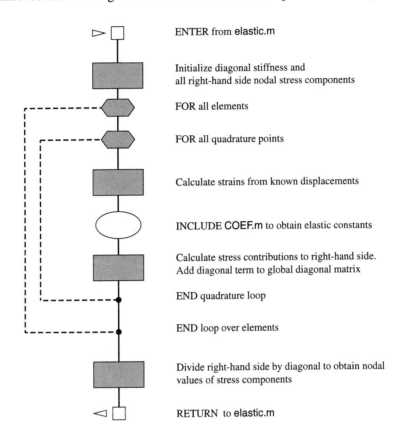

ENTER from elastic.m

Initialize diagonal stiffness and
all right-hand side nodal stress components

FOR all elements

FOR all quadrature points

Calculate strains from known displacements

INCLUDE COEF.m to obtain elastic constants

Calculate stress contributions to right-hand side.
Add diagonal term to global diagonal matrix

END quadrature loop

END loop over elements

Divide right-hand side by diagonal to obtain nodal
values of stress components

RETURN to elastic.m

11.10.6 Code: stress.m

<table>
<tr><td>

```
% ===================
% ROUTINE stress.m
% ===================

   NNPE2=2*NNPE;
   NUMEQ=NUMNP;

   for I=1:NUMNP
      SIGXX(I)=0.0;
      SIGXY(I)=0.0;
      SIGYY(I)=0.0;
      SIGZZ(I)=0.0;
      SGEFF(I)=0.0;
      SKD(I)=0.0;
   end

   for I=1:NUMEL

%   -----------------------------
%   Begin calculation of strain
%   and stress at each quadrature
%   point in each element.
%   -----------------------------
   JEND=NUMQPT(1);
   for J=1:JEND
      XJ        =0.0;
      YJ        =0.0;
      RJAC(1,1)=0.0;
      RJAC(1,2)=0.0;
      RJAC(2,1)=0.0;
      RJAC(2,2)=0.0;

      for K=1:NNPE
        NPK=NP(I,K);
        XJ = XJ + SF(1,K,J)*XORD(NPK);
        YJ = YJ + SF(1,K,J)*YORD(NPK);
        RJAC(1,1)=RJAC(1,1) ...
                  + SF(2,K,J)*XORD(NPK);
        RJAC(1,2)=RJAC(1,2) ...
                  + SF(3,K,J)*XORD(NPK);
        RJAC(2,1)=RJAC(2,1) ...
                  + SF(2,K,J)*YORD(NPK);
        RJAC(2,2)=RJAC(2,2) ...
                  + SF(3,K,J)*YORD(NPK);
      end
```

</td><td>

Routine stress.m
Called from elastic.m to calculate nodal
point stress components

NNPE2 = 2(number of nodes per
element)
NUMEQ = number of equations

Initialize right-hand-side vectors, one
vector for each component of stress.

SKD = vector of the lumped diagonal
stiffness matrix

Begin formulation of the global diagonal
matrix and right-hand-side vectors
element by element.

Begin Gaussian quadrature.

NUMQPT(1) = number of Gauss points
for element quadrature

Initialize quadrature point coordinates.

Initialize Jacobian matrix.

$$\left[J \frac{(x, y)}{(u, v)} \right] = 0$$

Calculate quadrature point values:

$$XJ = \lfloor N \rfloor \{X\} \qquad YJ = \lfloor N \rfloor \{Y\}$$

$RJAC(1,1) = \partial x / \partial u = \lfloor N_u \rfloor \{X\}$
$RJAC(1,2) = \partial x / \partial v = \lfloor N_v \rfloor \{X\}$
$RJAC(2,1) = \partial y / \partial u = \lfloor N_u \rfloor \{Y\}$
$RJAC(2,2) = \partial y / \partial v = \lfloor N_v \rfloor \{Y\}$

</td></tr>
</table>

```
        DETJ=RJAC(1,1)*RJAC(2,2) ...
                    - RJAC(2,1)*RJAC(1,2);
        if DETJ <= 0.0
            error
        end

        RJACI(1,1)=+RJAC(2,2)/DETJ;
        RJACI(1,2)=-RJAC(1,2)/DETJ;
        RJACI(2,1)=-RJAC(2,1)/DETJ;
        RJACI(2,2)=+RJAC(1,1)/DETJ;

        for K=1:NNPE
            DNDX(K)=RJACI(1,1)*SF(2,K,J) ...
                + RJACI(2,1)*SF(3,K,J);
            DNDY(K)=RJACI(1,2)*SF(2,K,J) ...
                + RJACI(2,2)*SF(3,K,J);
        end


%    ------------------------------------
%    Include user-written COEF.m code
%    ------------------------------------
        COEF


%    ------------------------------------
%    Determine R matrix corresponding
%    to analysis being performed
%    ------------------------------------
%    IPROG = 1      PLANE STRAIN ELASTICITY
%    IPROG = 2      PLANE STRESS ELASTICITY
%    ------------------------------------
        if IPROG == 1
            R(1,1)=2.0*G+LAMBDA;
            R(2,2)=R(1,1);
            R(3,3)=G;
            R(1,2)=LAMBDA;
            R(2,1)=LAMBDA;
            Mu = LAMBDA/(2*(LAMBDA+G));
        elseif IPROG == 2
            R(1,1)=4.0*G*(G+LAMBDA)/...
                        (2.0*G+LAMBDA);
            R(1,2)=2.0*G*LAMBDA/...
                        (2.0*G+LAMBDA);
            R(2,1)=R(1,2);
            R(2,2)=R(1,1);
            R(3,3)=G;
            Mu = LAMBDA/(2*(LAMBDA+G));
        end
```

Determinant of the Jacobian matrix:

$$\text{DETJ} = \|\text{RJAC}\|$$

If determinant is less than or equal to zero, there is an error in element mapping. Calculate inverse of the Jacobian.

$$[\text{RJACI}] = [\text{RJAC}]^{-1}$$

$$\left[J \frac{(u, v)}{(x, y)} \right] = \left[J \frac{(x, y)}{(u, v)} \right]^{-1}$$

Calculate gradient of shape functions with respect to global coordinates.

$$\begin{bmatrix} \lfloor \partial N / \partial x \rfloor \\ \lfloor \partial N / \partial y \rfloor \end{bmatrix} = J \frac{(u, v)}{(x, y)} \begin{bmatrix} \lfloor \partial N / \partial u \rfloor \\ \lfloor \partial N / \partial v \rfloor \end{bmatrix}$$

Include written COEF.m to obtain material constants.

G = shear modulus of elasticity
LAMBDA = λ, Lamé's second constant

Create [*R*] matrix according to type of analysis desired, i.e., plane strain or plane stress.

```
%       --------------------------------
%       CALCULATE CONTRIBUTION TO RHS
%       FOR STRESS
%       --------------------------------
        EPXX=0.0;
        EPXY=0.0;
        EPYY=0.0;
        for K=1:NNPE
            KY=NP(I,K)*2;
            KX=KY-1;
            UXK=U(KX);
            UYK=U(KY);
            EPXX=EPXX+DNDX(K)*UXK;
            EPYY=EPYY+DNDY(K)*UYK;
            EPXY=EPXY+(DNDY(K)*UXK+...
                      DNDX(K)*UYK)/2.0
        end


        SGXX=R(1,1)*EPXX+R(1,2)*EPYY;
        SGYY=R(2,1)*EPXX+R(2,2)*EPYY;
        SGXY=2.0*R(3,3)*EPXY;

        if IPROG == 1
            SGZZ=Mu*(SGXX+SGYY);
        else
            SGZZ=0;
        end

        for K=1:NNPE
            NPK=NP(I,K);
            XK=XORD(NPK);
            YK=YORD(NPK);
            wght=(XK-XJ)^2 + (YK-YJ)^2;
            wght=1.0/(sqrt(wght));
            SIGXX(NPK)=SIGXX(NPK)+wght*SGXX;
            SIGYY(NPK)=SIGYY(NPK)+wght*SGYY;
            SIGXY(NPK)=SIGXY(NPK)+wght*SGXY;
            SIGZZ(NPK)=SIGZZ(NPK)+wght*SGZZ;
            SKD(NPK)=SKD(NPK)+wght;
        end
    end
%   ---------    end of volume quadrature
    end
%   ---------    end of element loop
```

Initialize strain components.

KX and KY $=$ address for x and y components of displacement in the U array.

$$\epsilon_{xx} = \partial u / \partial x$$
$$\epsilon_{yy} = \partial v / \partial y$$
$$\epsilon_{xy} = (\partial u / \partial x + \partial u / \partial y)/2$$

$$\begin{Bmatrix} \sigma_{xx} \\ \sigma_{yy} \\ \sigma_{xy} \end{Bmatrix} = \begin{bmatrix} R_{11} & R_{12} & 0 \\ R_{12} & R_{22} & 0 \\ 0 & 0 & R_{33} \end{bmatrix} \begin{Bmatrix} \epsilon_{xx} \\ \epsilon_{yy} \\ \epsilon_{xy} \end{Bmatrix}$$

Calculate σ_{zz} according to plane strain or plane stress analysis.

Calculate weight to be given to quadrature values of stress components for each node in current element.

Here the weight given is the inverse of the distance from the quadrature point to the node.

(XJ, YJ) $=$ coordinates of quadrature point
(XK, YK) $=$ coordinates of nodal point

Distance $= \sqrt{(XK - XJ)^2 + (YK - YJ)^2}$

Weight $=$ 1/Distance

Sum weights to be the lumped diagonal terms.

```
% -----------------------------------
% Solve for nodal values of stress
% components by dividing by diagonal
% -----------------------------------
    for I=1:NUMNP
      SIGXX(I)=SIGXX(I)/SKD(I);
      SIGYY(I)=SIGYY(I)/SKD(I);
      SIGXY(I)=SIGXY(I)/SKD(I);
      SIGZZ(I)=SIGZZ(I)/SKD(I);
    end

% -------------------------------
%  Calculate nodal point value of
%  effective stress
% -------------------------------
    for I=1:NUMNP
        XX=SIGXX(I);
        YY=SIGYY(I);
        XY=SIGXY(I);
        ZZ=SIGZZ(I);
        XP=XORD(I);
        YP=YORD(I);

        SG=((XX-YY)^2 +(YY-ZZ)^2 ...
          +(ZZ-XX)^2 +6.0*XY^2);

        SGEFF(I)=sqrt(SG/2.0);
    end
```

Solve for weighted average of stress components by dividing by diagonal terms (the sum of the weights).

Calculate effective stress at each node:

$$\sigma_{\text{eff}} = \frac{1}{2}\Big[(\sigma_{xx} - \sigma_{yy})^2 + (\sigma_{yy} - \sigma_{zz})^2 \\ + (\sigma_{zz} - \sigma_{xx})^2 + 6(\sigma_{xy})^2\Big]^{1/2}$$

Return to elastic.m.

11.10.7 Test Problem. The problem selected to test elastic.m is a plate under uniform stress in the horizontal direction as shown.

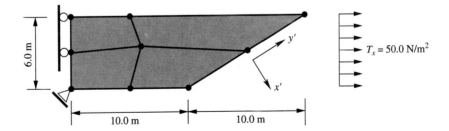

Note that on all surfaces, two boundary conditions are specified. It is important to realize that on the upper and lower surfaces, both components of traction are specified as zero. At the lower left corner, the vertical as well as the horizontal displacement is specified in order to prevent rigid body motion. To test the use of the local coordinate specification, the traction on the slanted, right surface will be specified in terms of the local coordinates shown rather than the more logical specification of $T_x = 50.0$ and $T_y = 0$. The input files and INCLUDE codes for this problem are as follows:

MESHo

```
9   % NUMNP
4   % NUMEL
4   % NNPE
```

NP

```
%-------------
   1   2   5   4
   2   3   6   5
   4   5   8   7
   5   6   9   8
%-------------
```

NODES

```
%   XORD    YORD    NPcode
%----------------------------
     0       0        3
     5       0        0
    10       0        7
     0       3        1
     6       3.5      0
    15       3        7
     0       6        1
     5       6        0
    20       6        7
%----------------------------
```

NWLD

```
%-----------
   2
   7
   6
   3
   4
   5
   8
   1
   9
   9   %   IB
%-----------
```

Above, NPcode has been defined as follows:

NPcode	Meaning
0	NPBC = 0
1	NPBC = 1
3	NPBC = 3
7	Node is on slanting surface

```
%------------------------------------
%          INITIAL.m
%
%  The following must be specified
%  if different from the default
%  values of zero.
%
%  NPBC, XBC, YBC, TX, TY, THETA
%------------------------------------

  for Ix=1:NUMNP
    if NPcode(Ix) == 1;
      NPBC(Ix)  = 1;
      XBC(Ix)   = 0.0;
      YBC(Ix)   = 0.0;
      TX(Ix)    = 0.0;
      TY(Ix)    = 0.0;
      THETA(Ix)= 0.0;
    elseif NPcode(Ix) == 3;
      NPBC(Ix)  = 3;
      XBC(Ix)   = 0.0;
      YBC(Ix)   = 0.0;
      TX(Ix)    = 0.0;
      TY(Ix)    = 0.0;
      THETA(Ix)= 0.3;
    elseif NPcode(Ix) == 7;
      angle     = atan(10./6.);
      NPBC(Ix)  = 0;
      XBC(Ix)   = 0.0;
      YBC(Ix)   = 0.0;
      TX(Ix)    = 300/sqrt(136)*cos(angle);
      TY(Ix)    = 300/sqrt(136)*sin(angle);
      THETA(Ix)= -angle;
      THETA(Ix)= 0.0;
    end
  end
```

INITIAL.m

This routine is included in elastic.m after the mesh data has been loaded. Hence, all variables associated with the mesh are available for use in assigning boundary values—specifically, the NPcode number given to each node.

NPcode: Each node has been assigned an identifying number by the user in the input file **NODES**. These numbers are used here to assign the proper boundary conditions. However, the coordinates and nodal point numbers can also be used to identify nodes for the assignment of boundary conditions.

NPBC: This number defines which variables are known (K) and unknown (U) at each node:

NPBC	0	1	2	3
U_x	U	K	U	K
U_y	U	U	K	K
F_x	K	U	K	U
F_y	K	K	U	U

XBC and YBC: The x and y components of nodal force (F_x and F_y) or displacement (U_x and U_y), whichever is the known value.

TX and TY: The x and y nodal components of traction

THETA: The counterclockwise rotation of the global coordinate axes into local axes

Note: XBC, YBC, TX, and TY components are to be given in terms of the local coordinates designated by the angle THETA.

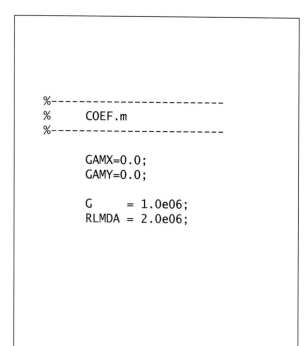

```
%-----------------------------
%      COEF.m
%-----------------------------

        GAMX=0.0;
        GAMY=0.0;

        G     = 1.0e06;
        RLMDA = 2.0e06;
```

COEF.m

This code is included in strain.m and stress.m within the quadrature loops for the volume integration to obtain values for

GAMX	x component of body force
GAMY	y component of body force
G	Shear modulus of elasticity
RLMDA	Lamé's constant (λ)

Variables available within the codes that are available for use in the definitions are

I	Element number
J	Quadrature point number
XJ	x coordinate of point
YJ	y coordinate of point
DNDX( )	$\partial N(\ )/\partial x$ at current point
DNDY( )	$\partial N(\ )/\partial y$ at current point
NPE( )	Array of node numbers for current element

With these auxiliary files, elastic.m, with the plane stress designation, gives $\sigma_{xx} = 50.0$ N/m^2, and u_x maximum $= 0.375E{-}03$ m. These values agree with the exact solution. When the x components of the nodal point displacements, UX, are extracted from the ASCII data file, SOLUTION, they produce uniformly spaced contours as shown:

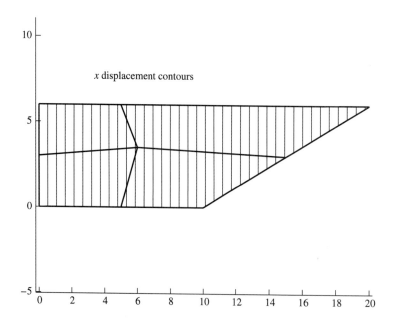

x displacement contours

EXERCISES

Study Problems

S1. Derive the differential equations of equilibrium based on a differential element.

S2. Prove that the stress tensor must be symmetric.

S3. Derive Eq. 11.4, the principle of virtual work.

S4. Explain why, for the first term in Eq. 11.4, the negative sign must be present in order for it to represent the work done by the internal forces.

S5. Why does the expression for virtual work represent the negative of the variation of potential energy?

S6. Verify Eqs. 11.11 and 11.12.

S7. Verify Eqs. 11.35, 11.36, and 11.37.

S8. Derive Eq. 11.39.

S9. Discuss the relationship between the reciprocal theorem of elasticity and the symmetry of the stiffness matrix.

S10. The uniqueness theorem of elasticity states that, for a given set of tractions, body forces, and specified displacements, there is only one solution to the equations for linear elasticity. Why, therefore, are finite element solutions not exact?

S11. Solve for the exact solution of the test problem using both plane stress and plane strain.

Numerical Experiments and Code Development

N1. Run the test problem using the components of traction specified in global coordinates rather than local coordinates.

N2. Develop a test problem of your own where both components of the body force are nonzero.

N3. Revise elastic.m to include axisymmetric analysis.

N4. Revise stress.m so that all nodal forces are calculated using the stiffness matrix and the displacements as found from strain.m. This will provide for the user the magnitude of unknown constraint forces.

N5. Elastic materials, when subjected to an increase in temperature, undergo a thermal expansion that creates a state of isotropic thermal strain given by

$$
\begin{bmatrix} \epsilon_{xx} & \epsilon_{xy} & \epsilon_{xz} \\ \epsilon_{yx} & \epsilon_{yy} & \epsilon_{yz} \\ \epsilon_{zx} & \epsilon_{zy} & \epsilon_{zz} \end{bmatrix} = \begin{bmatrix} \alpha\Delta T & 0 & 0 \\ 0 & \alpha\Delta T & 0 \\ 0 & 0 & \alpha\Delta T \end{bmatrix}
$$

where α is the coefficient of thermal expansion and ΔT is the change in temperature. Thus, at any given point, the elastic strains given in Eq. 11.1 are the total strains less the thermal strains given above. Symbolically, we can write

$$ \{\sigma\} = [R]\{\epsilon - \alpha\Delta T\} $$

where ϵ represents the total strains. Note that the terms represented by $\alpha\Delta T$ are only for the normal components, not the shear components. If this formulation is substituted into the weak form of our governing equation, a new term will appear, equal to

$$\int_{V_e} \lfloor \delta\epsilon \rfloor [R]\{\alpha\Delta T\} \ dV$$

which can be placed on the right-hand side of our finite element equations as a force vector. This new force vector will be self-equilibrating and will create strains (and hence thermal stresses) resulting from changes in temperature. Revise both strain.m and stress.m to accommodate thermal expansion.

N6. For our plane strain and plane stress formulations, we assumed $\epsilon_{zz} = 0$ (and hence $\delta\epsilon_{zz} = 0$) and $\sigma_{zz} = 0$, respectively. For both cases $\delta\epsilon_{zz}\sigma zz = 0$, which resulted in identical expressions for virtual work. However, for long cylinders with no end constraints, neither of the above assumptions is true. We can deal with these situations by first solving the problem under the assumption of plane strain.

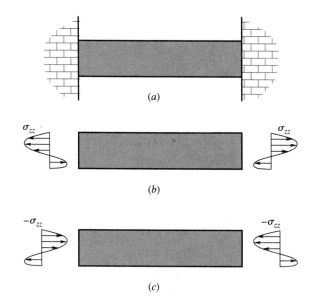

This is illustrated in part a of the accompanying figure. This solution creates a $\sigma_{zz}(x, y)$ value that will be the same for all cross-sections, including those at both ends of the cylinder. This is illustrated in part b of the figure. If we now solve the problem of an identical cylinder loaded only at its ends with negative σ_{zz} stresses (figure part c), we can superimpose this solution onto our plane strain solution to obtain our desired solution (note that after superposition, $\sigma_{zz} = 0$ on each end). This is referred to as the generalized plane strain problem.

We must now determine the solution to the problem of a long cylindrical rod, subjected to axial loads at its ends corresponding to the stress distribution $-\sigma_{zz}$. For this we make use of St. Venant's principle, which states that the stress distribution at a cross-section in the rod that is far removed from the ends will be the same for all statically equivalent loadings on the ends of the rod. Fortunately, we have a very simple solution from elementary strength of materials for such a problem. We need only calculate the bending moments about the principal axes of the cross-section caused by σ_{xx} and the axial force caused by it. Once this is obtained, application of the flexure formula and the P/A formula will give us σ_{zz} due to this loading.

Create a postprocessor code that takes the nodal point component for σ_{zz} and does this. The following outline can be used:

(a) Load the mesh data and σ_{zz} from your plane strain analysis.

(b) Begin a numerical integration to determine centroid, I_{xx}, I_{yy}, I_{xy}, total axial force, M_x, and M_y. Let each element be the ΔA for the numerical integrations.

(c) Use the parallel axis theorem to determine the moment of inertia around the centroidal axis.

(d) Determine the principal axes of the cross-section.

(e) Determine the moment about the principal axes.

(f) Use the flexure formula and P/A to determine σ_{xx} due to the axial loading.

(g) Add these stresses to those found using the plane strain analysis to obtain σ_{xx} for the generalized plane strain problem.

Projects

P1. For the two rods shown (select your own dimensions), loaded under uniform tension, determine the stress distribution around the notches. Assume plane stress and make use of the two lines of symmetry.

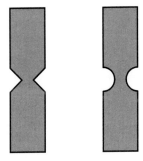

P2. Rock and other brittle materials can be tested for tensile strength by loading a cylinder as shown. The procedure is known as the Brazilian tensile test. When the specimen cracks, the tensile strength is taken to be

$$\sigma_t = \frac{P}{\pi R L}$$

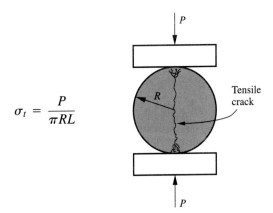

where R = radius, L = length of specimen, and P = total load. Use elastic.m to evaluate the accuracy of this equation. Compare the plane strain and plane stress results.

P3. The following photoelastic images show fringe patterns corresponding to regions where the difference in the principal stresses is constant (taken from Frocht, 1961). The patterns, therefore, correspond to regions of constant maximum shear stress in the x-y plane. Part a of the accompanying figure corresponds to the Brazilian tensile test, part b is similar but is a ring rather than a disc, part c is a split ring loaded at notches, part d is a disc loaded at the top and supported at two points, and part e shows the stress distribution around a hole in a plate loaded in the vertical direction. For each, compare the results from a similar finite element plane stress analysis using contours of constant effective stress. These will be similar to constant maximum shear stress, but you may want to revise stress.m to calculate maximum shear stress as well as effective stress.

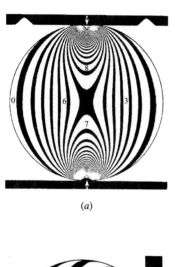

(a)

(b)

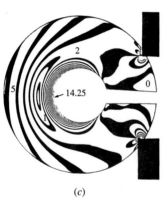

(c)

(d)

(e)

P4. Use the revision of elastic.m for thermal strains as suggested in Exercise N5 to obtain the solution to the following problem.

Two pieces of identical material, shown in the figure, are fitted together by cooling the inner object and heating the outer object by the same number of degrees relative to the ambient temperature. At these temperatures, the two pieces fit snugly. After both pieces have returned to ambient temperature, calculate the thermal stresses in the assembly.

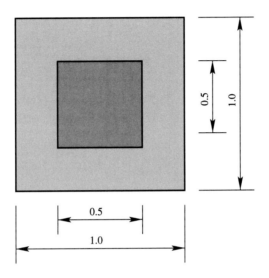

For your analysis, use the following values:

G	λ	α	ΔT
1.0	1.4	1.0	1.0

The value given for ΔT is the amount (plus or minus) that each piece has been changed from ambient temperature. Note that the values for G and λ give a value for Poisson's ratio equal to 0.292. For this linear analysis, the stress will given in terms of G, per $\alpha \Delta T$, where $\alpha \Delta T$ is dimensionless.

Use the maximum value of effective stress found from these parametric values to determine what value of ΔT would cause the material to yield if it were structural steel. (See figures on next page.)

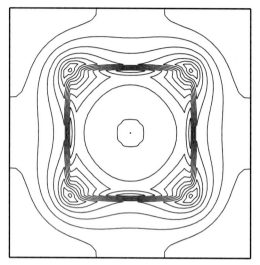

Contours of effective stress.

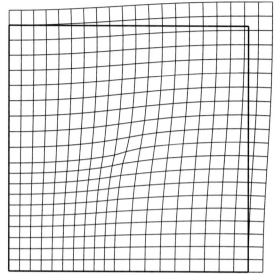

Deformed mesh (first quadrant).

12

HIGHER-ORDER EQUATIONS

Thus far in this text we have restricted the finite element method to the solution of second-order differential equations. This allows the analysis of a large number of problems of practical importance. However, some phenomena are best described by higher-order differential equations. Examples of these appear in structural mechanics and include the deflection of beams and plates. Although the topic of higher-order partial differential equations, such as those used to describe the deflection of plates, is beyond the scope of this book, we present in this chapter a finite element approximation of a fourth-order ordinary differential equation that describes the deflection of a beam on an elastic foundation. Not only will this serve as an introduction to the analysis of higher-order equations, but it will be useful in itself; such problems are not easily solved.

12.1 BEAMS ON ELASTIC FOUNDATIONS

Figure 12.1 illustrates the type of problem we are considering: a beam supported on a foundation of elastic springs. The springs are to be considered continuous along the length of the beam, but discontinuous from each other. Thus, the deflection of the foundation and the support it gives to the beam at any given point are independent of the deflection at neighboring points.[1] This mathematical model for foundation support is referred to as a *Winkler foundation*, named for the person who first proposed it.

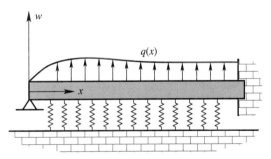

Figure 12.1. Beam on an elastic foundation.

With the assumption of a Winkler foundation along with the usual assumptions associated with elementary beam theory, equilibrium of a differential length of the beam requires

$$\frac{d^2}{dx^2}\left(EI\frac{d^2w}{dx^2}\right) + kw - q = 0$$

---

[1] One might think of advertisements for certain mattresses that boast similar properties.

where

w = deflection $[L]$
x = coordinate along the beam $[L]$
E = modulus of elasticity for the beam $[F/L^2]$
I = moment of inertia of the beam's cross-section $[L^4]$
k = foundation modulus $[F/L^2]$
q = loading on the beam. $[F/L]$

What distinguishes this equation from all equations we have considered up to this point is the fact that it implies that the fourth derivative of w exists;[2] hence, it is necessary that any approximation to this equation have continuous third derivatives. This requirement places a strong demand on all such approximations; thus, we follow our usual procedure for writing the equation in a weaker form. For this we use our standard variational notation and write

$$\int_0^L \delta w \left[\frac{d^2}{dx^2} \left(EI \frac{d^2 w}{dx^2} \right) + kw - q \right] dx = 0$$

$$\int_0^L \left\{ \frac{d}{dx} \left[\delta w \frac{d}{dx} \left(EI \frac{d^2 w}{dx^2} \right) \right] - \left[\frac{d \, \delta w}{dx} \frac{d}{dx} \left(EI \frac{d^2 w}{dx^2} \right) \right] + \delta w \, kw - \delta w \, q \right\} dx = 0$$

$$\left[\delta w \frac{d}{dx} \left(EI \frac{d^2 w}{dx^2} \right) \right]_0^L - \int_0^L \left\{ \left[\frac{d \, \delta w}{dx} \frac{d}{dx} \left(EI \frac{d^2 w}{dx^2} \right) \right] - \delta w \, kw + \delta w \, q \right\} dx = 0 \qquad (12.1)$$

$$\left[\delta w \frac{d}{dx} \left(EI \frac{d^2 w}{dx^2} \right) \right]_0^L - \int_0^L \left\{ \frac{d}{dx} \left[\frac{d \, \delta w}{dx} EI \frac{d^2 w}{dx^2} \right] - \left[\frac{d^2 \delta w}{dx^2} EI \frac{d^2 w}{dx^2} \right] - \delta w \, kw + \delta w \, q \right\} dx = 0$$

$$\left[\delta w \frac{d}{dx} \left(EI \frac{d^2 w}{dx^2} \right) \right]_0^L - \left[\frac{d \, \delta w}{dx} EI \frac{d^2 w}{dx^2} \right]_0^L + \int_0^L \left[\frac{d^2 \delta w}{dx^2} EI \frac{d^2 w}{dx^2} + \delta w \, kw - \delta w \, q \right] dx = 0$$

At this point we have integrated by parts twice and reduced the formulation to second order. Note that another integration by parts would result in the third derivative of δw appearing, thus accomplishing nothing toward our goal of having a weaker formulation. We are therefore stuck with this as the weakest formulation we can obtain for our problem. It is this fact that places much stricter demands on our finite element approximations than were placed on them in our previous problems. How we meet these demands will be presented in the next section.

The boundary conditions implied in Eq. 12.1 can be expressed in terms of the more usual notations for shear and moments by using

---

[2]At least for beams with continuous EI, which all beams have at some points!

$$M = EI\frac{d^2w}{dx^2}$$

$$V = \frac{d}{dx}\left(EI\frac{d^2w}{dx^2}\right)$$

We can now write Eq. 12.1 as

$$\delta w_L\, V_L - \delta w_0\, V_0 - \left(\frac{d\delta w}{dx}\right)_L M_L + \left(\frac{d\delta w}{dx}\right)_0 M_0 + \int_0^L \left\{\left[\frac{d^2\delta w}{dx^2}EI\frac{d^2w}{dx^2}\right] + \delta w\, kw - \delta w\, q\right\} dx = 0 \quad (12.2)$$

Note that the weak form of our equation gives us the boundary conditions that must be imposed at the ends of the beam:

V known and w unknown, hence δw arbitrary
or
V unknown and w known, hence $\delta w = 0$

and

M known and w' unknown, hence $\delta w'$ arbitrary
or
M unknown and w' known, hence $\delta w' = 0$

Inspection of Eq. 12.2 shows that it is the negative of the virtual work performed by all forces in the system during a virtual displacement δw. Thus, if all forces are constant, it represents the variation of the total potential energy of the system given by the functional

$$PE = \int_0^L \left(\frac{1}{2}EI\left(\frac{d^2w}{dx^2}\right)^2 + \frac{1}{2}kw^2 - wq\right) dx + w_L V_L - w_0 V_0 - \left(\frac{dw}{dx}\right)_L M_L + \left(\frac{dw}{dx}\right)_0 M_0 \quad (12.3)$$

12.2 FINITE ELEMENT APPROXIMATION

Because the weak form given by Eq. 12.2 requires the existence of the second derivatives of both the trial function and the test function, these functions must be continuous and have continuous first derivatives. Hence, it is necessary that a finite element approximation exhibit the same smoothness at all nodes. This can be accomplished by using Hermite cubics. These functions are piecewise cubics defined by the nodal values of both the function and its derivative; thus, they serve our purpose perfectly. With these cubics, our finite element approximations will be given in terms of the nodal values for w and w' as indicated in Fig. 12.2. Clearly, because adjacent elements have the same nodal values for both w and w', these quantities will be continuous at all nodes, thus satisfying the smoothness requirement.

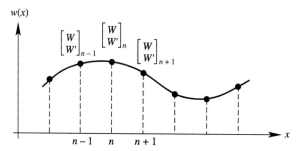

Figure 12.2. Piecewise Hermite cubic approximating functions.

12.2.1 Hermite Cubics. We now show how Hermite cubics can be used as the basis of our finite element approximation. In doing so, we will use the same technique as we did for program ode2.m and define our approximation for a single element using its Gaussian coordinates. Once this is done, we will consider the mapping of this approximation to the x coordinate associated with our problem.

Within a single element we wish to define our approximation in terms of the nodal values for w and dw/du at each of its ends. These four nodal parameters will each have an associated shape function, and our approximation will be written as

$$w(u) = \left\lfloor N_1(u) \quad N_2(u) \quad N_3(u) \quad N_4(u) \right\rfloor \begin{Bmatrix} W_a \\ dW_a/du \\ W_b \\ dW_b/du \end{Bmatrix} \tag{12.4}$$

where a and b represent the left and right nodes.

Here, however, we note an important deviation from previous shape functions. Whereas all our previous functions have been dimensionless, that is no longer the case. A dimensional analysis of Eq. 12.4 reveals that

$$w(u) \quad = \quad N_1 W_a \quad + \quad N_2 W_a' \quad + \quad N_3 W_b \quad + \quad N_4 W_b'$$

(12.5)

| L | $1\,L$ | $L\dfrac{L}{L}$ | $1\,L$ | $L\dfrac{L}{L}$ |

Hence, N_2 and N_4 have the dimension of length. To keep track of this dimensionality in the following equations, we will assume our element has the coordinates of $\pm a$ rather than ± 1 associated with the Gaussian coordinates.

Because our approximation is a cubic, each shape function will be a cubic having a value equal to unity for its corresponding parameter and zero for the other three parameters. Knowing this, it is a simple exercise to arrive at the following equations for the four shape functions:

$$N_1 = \left(\frac{1}{4}\right)\left(\frac{u}{a} + 2\right)\left(\frac{u}{a} - 1\right)^2$$

$$N_2 = \left(\frac{a}{4}\right)\left(\frac{u}{a} + 1\right)\left(\frac{u}{a} - 1\right)^2$$

$$N_2 = \left(\frac{1}{4}\right)\left(\frac{u}{a} - 2\right)\left(\frac{u}{a} + 1\right)^2 \qquad (12.6)$$

$$N_2 = \left(\frac{a}{4}\right)\left(\frac{u}{a} - 1\right)\left(\frac{u}{a} + 1\right)^2$$

Clearly, both N_2 and N_4 have the dimension associated with a, which is length. Keeping this in mind, we now write these equations to correspond to the Gaussian coordinate where $a = 1$:

$$N_1 = (1/4)(u + 2)(u - 1)^2$$

$$N_2 = (1/4)(u + 1)(u - 1)^2$$

$$N_3 = -(1/4)(u - 2)(u + 1)^2 \qquad (12.7)$$

$$N_4 = (1/4)(u - 1)(u + 1)^2$$

These functions are illustrated in Fig. 12.3.

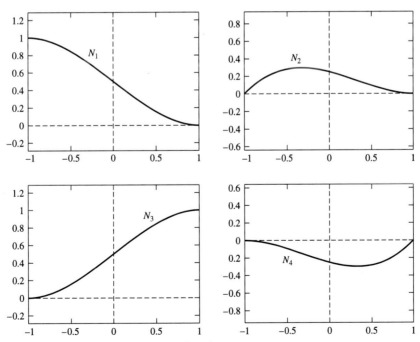

Figure 12.3. Hermite cubic shape functions.

We will also have need for the first and second derivatives of these shape functions. They are

$$\frac{dN_1}{du} = (1/4)(3u^2 - 3)$$

$$\frac{dN_2}{du} = (1/4)(3u^2 - 2u - 1)$$

$$\frac{dN_3}{du} = -(1/4)(3u^2 - 3)$$

$$\frac{dN_3}{du} = (1/4)(3u^2 + 2u - 1)$$

(12.8)

and

$$\frac{d^2N_1}{du^2} = (1/4)(6u)$$

$$\frac{d^2N_2}{du^2} = (1/4)(6u - 2)$$

$$\frac{d^2N_3}{du^2} = -(1/4)(6u)$$

$$\frac{d^2N_3}{du^2} = (1/4)(6u + 2)$$

(12.9)

Having defined our shape functions with respect to the Gaussian coordinate, u, we now consider how they are mapped to the x axis.

12.2.2 Mapping the Hermite Cubics. If the mapping of points along the Gaussian axis, u, to points along the x axis is single-valued and continuous, we can express it and its unique inverse as

$$x = x(u) \quad \text{and} \quad u = u(x)$$

(12.10)

For any such mapping, the value of the shape functions at any point along the x axis can be determined by simple substitution. Thus, we can write

$$N_1(x) = N_1\big(u(x)\big)$$

$$N_3(x) = N_3\big(u(x)\big)$$

(12.11)

These two shape functions give us $w(x)$ in terms of values of W at the ends of the element. However, if we define N_2 and N_4 in this manner, they will correspond to values of $w(x)$ in terms of dw/du at each of the two ends.

Clearly, this is not what we want; rather, we want these two shape functions to correspond to $w(x)$ in terms of dw/dx at each end of the element. Thus, we write

$$N_2(u)\frac{dw}{du} = N_2\big(u(x)\big)\frac{dw}{du}$$

$$= N_2\big(u(x)\big)\frac{dx}{du}\frac{dw}{dx}$$

(12.12)

and define

$$N_2(x) = N_2\big(u(x)\big)\frac{dx}{du}$$

(12.13)

and similarly,

$$N_4(x) = N_4\big(u(x)\big)\frac{dx}{du}$$

(12.14)

These four functions can now be used to define our approximation of $w(x)$ in terms of the nodal values of w and dw/dx, which we write as

$$w(x) = \lfloor\; N_1(x) \quad N_2(x) \quad N_3(x) \quad N_4(x) \;\rfloor \begin{Bmatrix} W_a \\ dW_a/dx \\ W_b \\ dW_b/dx \end{Bmatrix}$$

(12.15)

We next consider the derivatives of w with respect to x. These are

$$\frac{dw}{dx} = \lfloor dN/dx\rfloor\{W\}$$

$$\frac{d^2w}{dx^2} = \lfloor d^2N/dx^2\rfloor\{W\}$$

(12.16)

where we use the symbol $\{W\}$ to represent the array of nodal values of both w and dw/dx. Because our shape functions are defined in terms of u, we must use the chain rule of differentiation to obtain derivatives with respect to x. Again, for N_1 and N_3, this is straightforward and we have

$$\frac{dN_1}{dx} = \left(\frac{dN_1}{du}\right)\frac{du}{dx}$$

$$\frac{dN_3}{dx} = \left(\frac{dN_3}{du}\right)\frac{du}{dx}$$

(12.17)

and

$$\frac{d^2N_1}{dx^2} = \left(\frac{d^2N_1}{du^2}\right)\left(\frac{du}{dx}\right)^2 + \left(\frac{dN_1}{du}\right)\frac{d^2u}{dx^2}$$

$$\frac{d^2N_3}{dx^2} = \left(\frac{d^2N_3}{du^2}\right)\left(\frac{du}{dx}\right)^2 + \left(\frac{dN_3}{du}\right)\frac{d^2u}{dx^2}$$

(12.18)

For N_2 and N_4, things are a bit more tedious but nevertheless straightforward. We have

$$\frac{dN_2}{dx} = \frac{d}{dx}\left[N_2\frac{dx}{du}\right]$$

$$= \left(\frac{dN_2}{du}\right)\frac{du}{dx}\frac{dx}{du} + N_2\frac{d}{dx}\left(\frac{dx}{du}\right)$$

$$= \left(\frac{dN_2}{du}\right)\frac{du}{dx}\frac{dx}{du}$$

(12.19)

$$\frac{dN_4}{dx} = \frac{d}{dx}\left[N_4\frac{dx}{du}\right]$$

$$= \left(\frac{dN_4}{du}\right)\frac{du}{dx}\frac{dx}{du}$$

and

$$\frac{d^2N_2}{dx^2} = \frac{d}{dx}\left[\frac{dN_2}{dx}\right]$$

$$= \left(\frac{d^2N_2}{du^2}\right)\frac{du}{dx}\frac{du}{dx}\frac{dx}{du} + \left(\frac{dN_2}{du}\right)\frac{d^2u}{dx^2}\frac{dx}{du} + \left(\frac{dN_2}{du}\right)\frac{du}{dx}\frac{d}{dx}\left(\frac{dx}{du}\right)$$

$$= \left(\frac{d^2N_2}{du^2}\right)\left(\frac{du}{dx}\right)^2\frac{dx}{du} + \left(\frac{dN_2}{du}\right)\frac{d^2u}{dx^2}\frac{dx}{du}$$

(12.20)

$$\frac{d^2N_4}{dx^2} = \frac{d}{dx}\left[\frac{dN_4}{dx}\right]$$

$$= \left(\frac{d^2N_4}{du^2}\right)\left(\frac{du}{dx}\right)^2\frac{dx}{du} + \left(\frac{dN_4}{du}\right)\frac{d^2u}{dx^2}\frac{dx}{du}$$

If we now let $x(u)$ be the same linear transformation that we used for program ode2.m, we have

$$x = \left(\frac{a+b}{2}\right) + \left(\frac{b-a}{2}\right)u$$

$$dx = \left(\frac{b-a}{2}\right)du \tag{12.21}$$

$$du = \left(\frac{2}{b-a}\right)dx$$

Note that within an element both dx/du and du/dx are constant; thus, the second derivatives are zero. For this linear transformation we therefore have

Left node	*Right node*

$$N_1(x) = N_1 \qquad\qquad N_3(x) = N_3$$

$$N_2(x) = N_2\,\frac{dx}{du} \qquad\qquad N_4(x) = N_4\,\frac{dx}{du}$$

$$\frac{dN_1}{dx} = \frac{dN_1}{du}\frac{du}{dx} \qquad\qquad \frac{dN_3}{dx} = \frac{dN_3}{du}\frac{du}{dx}$$

$$\frac{dN_2}{dx} = \frac{dN_2}{du}\frac{du}{dx}\frac{dx}{du} \qquad\qquad \frac{dN_4}{dx} = \frac{dN_4}{du}\frac{du}{dx}\frac{dx}{du} \tag{12.22}$$

$$\frac{d^2N_1}{dx^2} = \frac{d^2N_1}{du^2}\left(\frac{du}{dx}\right)^2 \qquad\qquad \frac{d^2N_3}{dx^2} = \frac{d^2N_3}{du^2}\left(\frac{du}{dx}\right)^2$$

$$\frac{d^2N_2}{dx^2} = \frac{d^2N_2}{du^2}\left(\frac{du}{dx}\right)^2\frac{dx}{du} \qquad\qquad \frac{d^4N_4}{dx^2} = \frac{d^2N_4}{du^2}\left(\frac{du}{dx}\right)^2\frac{dx}{du}$$

Additional simplification could be made by noting that $(du/dx)(dx/du) = 1$. However, when this is done, the resulting equations can be misleading; hence, we write them as above for clarity.

These expressions can now be used to give us

$$w = \lfloor N \rfloor\{W\} \qquad \delta w = \lfloor N \rfloor\{\delta W\}$$

$$w' = \lfloor N' \rfloor\{W\} \qquad \delta w' = \lfloor N' \rfloor\{\delta W\} \tag{12.23}$$

$$w'' = \lfloor N'' \rfloor\{W\} \qquad \delta w'' = \lfloor N'' \rfloor\{\delta W\}$$

where the primes indicate differentiation with respect to x.

We now substitute these approximations into the weak form of our governing equation to obtain a finite element approximation to our problem. As before, we perform the integration element by element. The resulting element integrations are

$$
\int_0^{L_e} \frac{d^2 \delta w}{dx^2} EI \frac{d^2 w}{dx^2} \, dx = \int_0^{L_e} \lfloor \delta W \rfloor \{N''\} EI \lfloor N'' \rfloor \{W\} \, dx
$$

$$
= \lfloor \delta W \rfloor \left[\int_0^{L_e} \{N''\} EI \lfloor N'' \rfloor \, dx \right] \{W\}
\tag{12.24}
$$

$$
= \lfloor \delta W \rfloor [S_1] \{W\}
$$

$$
\int_0^{L_e} \delta w \, kw \, dx = \int_0^{L_e} \lfloor \delta W \rfloor \{N\} \, k \lfloor N \rfloor \{W\} \, dx
$$

$$
= \lfloor \delta W \rfloor \left[\int_0^{L_e} \{N\} \, k \lfloor N \rfloor \, dx \right] \{W\}
\tag{12.25}
$$

$$
= \lfloor \delta W \rfloor [S_2] \{W\}
$$

$$
\int_0^{L_e} \delta w \, q \, dx = \int_0^{L_e} \lfloor \delta W \rfloor \{N\} \, q \, dx
$$

$$
= \lfloor \delta W \rfloor \left\{ \int_0^{L_e} \{N\} \, q \, dx \right\}
\tag{12.26}
$$

$$
= \lfloor \delta W \rfloor \{q\}
$$

12.2.3 Summary of Equations

$$
[S_1]_e = \int_0^{L_e} \{N''\}^T EI \lfloor N'' \rfloor \, dx
$$

$$
[S_2]_e = \int_0^{L_e} \{N\}^T k \lfloor N \rfloor \, dx
\tag{12.27}
$$

$$
\{f\}_e = \int_0^{L_e} \{N\} \, q \, dx
$$

12.3 DIMENSIONAL ANALYSIS OF PROBLEM

In the previous chapters we used the concepts that underlie dimensional analysis without going into significant detail. This was possible because the application of these concepts in the analysis of our previous problems was fairly transparent. However, for beams on elastic foundations, these ideas become less transparent. For that reason, we pause here to present more of the details behind the method.

The fundamental goal is to remove from our problem all arbitrary units of measure such as meters and newtons. We will discover that in doing so, we will also reduce the number of independent parameters necessary to describe our problem.

For beams on elastic foundations, the only independent dimensions needed to describe the problem are length and force. Because we are free to select any units we wish for these dimensions, we can select ones that are natural to the problem. Hence, we will use the length of the beam, L, to be our unit of length and EI/L^2 to be our unit of force. If EI is not constant, we will select some representative value for it, such as its maximum value or its minimum value.

When all parameters and variables in a governing equation are given in terms of units such as these, the equation is said to be nondimensionalized. The reasoning behind this terminology is that all quantities are now measured in terms of their ratios with other quantities associated with the problem. In actuality, any unit is used simply to measure a quantity by its ratio with some standard that society has agreed to use.

To illustrate these concepts, consider all of the variables and parameters necessary to completely describe our problem. Each can be represented as a dimensionless number representing its ratio with some combination of EI and L. This is shown in the following table (for simplicity, consider all parameters as having constant values).

Original Variable	Original Dimensions	Dimensionless Variable (Ratio)
x	L	$\mathbf{x} = (x/L)$
k	F/L^2	$\mathbf{k} = k/(EI/L^4)$
q	F/L	$\mathbf{q} = q/(EI/L^3)$
w	L	$\mathbf{w} = w/(L)$
L	L	$\mathbf{L} = L/(L) = 1.0$
EI	FL^2	$\mathbf{EI} = EI/(EI) = 1.0$

Notice that regardless of the length of the beam, $\mathbf{L}$ is unity. The same is true of $\mathbf{EI}$. For this reason these two quantities will not appear explicitly in our problem statement; their value is implied in the measure of all of the other variables and parameters. Our problem can now be written as

$$\frac{d^4\mathbf{w}}{d\mathbf{x}^4} + \mathbf{kw} - \mathbf{q} = 0 \tag{12.28}$$

with appropriate boundary conditions at $\mathbf{x} = 0$ and $\mathbf{x} = 1$.

To complete this illustration, suppose we solved the above equation for $\mathbf{q} = 1.0$ and found that the largest value for $\mathbf{w}$ was 0.00253. Because the problem is linear with respect to $\mathbf{q}$, the deflection in terms of any other value of $\mathbf{q}$ would be

$$\mathbf{w} = 0.00253\mathbf{q}$$

If we now write the above nondimensional variables in terms of their defined ratios, we will obtain

$$\frac{w}{L} = 0.00253 \frac{q}{(EI/L^3)}$$

or

$$w = 0.0253 \frac{qL^4}{EI} \tag{12.29}$$

Thus, we have the deflection in terms of q, L, and EI given in whatever units we choose to use. Note also that this equation would hold for any *similar* beam, that is, a beam whose nondimensional parameters

$$\frac{kL^4}{EI} \quad \text{and} \quad \frac{qL^3}{EI}$$

are the same as those used in the original equation. This is the reason for the terminology *similitude* and *similarity analysis*.[3]

Such analyses are important for experimental investigations because they give the rules for scaling models: All variables must be scaled so that they create the same values for the non-dimensional terms as do the parameters associated with the prototype. The method, or theory, is also important for analytical investigations because (1) it provides a method for obtaining solutions that are independent of a particular set of units, and (2) the number of parameters necessary to specify an analysis will be reduced by the number of independent dimensions used to describe the problem.[4]

Note that in Eq. 12.28, the parameter **k** is the only physical parameter necessary to describe the beam-foundation system. Thus, for a given loading, **q**, and the same boundary conditions, any beam-foundation system that has the same **k** value will have exactly the same solution.

This parameter is similar to another parameter associated with the general solution for a beam with constant values for EI and k. That parameter is usually given the symbol β and is defined as

$$\beta = \left[\frac{k}{4EI} \right]^{1/4} \tag{12.30}$$

which has the dimensions of $1/L$. The use of the general solution for a particular problem requires satisfying the boundary conditions through specification of the parameters associated with the general solution. This is not always a simple task; hence, tables for specific solutions are made available in books. These solutions are arranged using values of β and are classified as solutions for short, intermediate, or long beams. This terminology is used whether or not the beam is actually short or long with respect to its depth or some other measure. A "long" beam means that its deflection will be prevented more by the foundation than by its stiffness. A "short" beam means that its deflection is prevented more by its stiffness than by the foundation. Hence, a "short" beam can be made long by increasing the foundation modulus, and a "long" beam can be made short by decreasing the foundation modulus.

---

[3]In the literature on dimensional analysis, the nondimensional parameters are referred to as *Pi terms*, and the fact that the number of independent parameters can be reduced by the number of independent dimensions necessary to describe them is known as the *Buckingham Pi theorem*.

[4]The method also plays an important role in approximation theory because it allows numerical comparison of non-dimensional terms in an equation. Such comparisons can indicate that one or more of the terms can be neglected, thus simplifying the equation.

12.4 PROGRAMMING PRELIMINARIES

Most of the programming techniques we will use for our new program are similar to those used in program ode2.m. The primary difference is that we now have two parameters per node rather than one.

12.4.1 The Global Matrices. The 4×4 element stiffness matrices will be assembled into the global stiffness matrix as shown in Fig. 12.4. Here it has been assumed that the mesh consists of three elements. The notations f and m on the right-hand side represent the discretized distributed loading corresponding to the shape functions for unit displacements and unit rotations at the respective nodes. They are equivalent to the point forces and moments that would produce the same work as the distributed load during a virtual displacement and a virtual rotation of a node. The corresponding boundary shears and moments are also shown for the first and last nodes of the mesh. Note that the bandwidth will be equal to 4 when stored as a symmetric banded matrix.

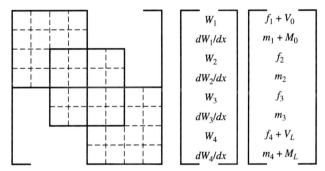

Figure 12.4. Schematic of the assembly of a three-element mesh.

12.4.2 Quadrature. Because our shape functions are cubic polynomials, the integrands associated with the $[S_2]$ matrix, when k is constant, are sixth-degree polynomials. Thus, it is necessary to use four quadrature points, which will allow a seventh-degree polynomial to be integrated exactly. This will also produce an exact integration if k varies linearly, or a good approximation if it can be accurately approximated within an element as a linear function. The Gauss points and weights, for four points, are

Coordinate	Weight
−0.861136311594953	0.347854845137454
−0.339981043584856	0.652145154862546
0.339981043584856	0.652145154862546
0.861136311594953	0.347854845137454

12.4.3 Boundary Conditions. Because both the deflection, w, and the slope, dw/dx, are nodal point variables, there are four possible conditions that can be given at the boundaries (or any other node). The following table defines the NPBC values for each condition

NPBC	w	dw/dx	V	M
0	U	U	K	K
1	K	U	U	K
2	U	K	K	U
3	K	K	U	U

where K = known and U = unknown.

12.4.4 COEF.m. The user-written INCLUDE file COEF.m must define EI, k, and dEI/dx. The last item is necessary in order to evaluate the shear, V (see below).

Some beam-foundation systems have pins in the beam that permit the transfer of shear but not moment. These pins can be simulated by defining a very short element at the pin's location, and then, within COEF.m,

1. Assign values for EI and k equal to zero for this element.
2. Set the $[S_1]$ matrix equal to

$$\begin{bmatrix} +B & 0 & -B & 0 \\ 0 & 0 & 0 & 0 \\ -B & 0 & +B & 0 \\ 0 & 0 & 0 & 0 \end{bmatrix}$$

where B stands for a numerical value that is large compared with other values that appear in the stiffness matrix. Thus, B serves as a penalty function that forces the deflections associated with the end nodes of this element to be equal. On the other hand, their slopes will be decoupled because neither $[S_1]$ nor $[S_2]$ will have terms that connect them.

Because COEF.m is called at each quadrature point, the above specification will be made as many times as there are quadrature points per element. This, however, is perhaps less expensive than building in some type of logic that determines if the specification has yet been made. The user can choose which approach to use.

Finally, it should be noted that the zero value given for EI will signal program winkler.m not to include this "fictitious" element in the calculation for moment and shear (see below).

12.4.5 Output. Of particular interest to engineers are the values for shear and moment in the beam. Their values, in terms of the deflection, are

$$M = EI \frac{d^2w}{dx^2}$$

$$V = \frac{d}{dx}\left(EI \frac{d^2w}{dx^2}\right) \tag{12.31}$$

$$= \frac{dEI}{dx}\frac{d^2w}{dx^2} + EI \frac{d^3w}{dx^3}$$

For beams of constant EI, the first term in the equation for V will be zero. However, its inclusion in the code will allow us to calculate the correct value for shear in beams with varying cross-sectional properties.

Because the equations for both M and V contain derivatives of w beyond the first, these quantities will be discontinuous at nodes; hence, we will calculate these values at the center of the elements and report all values with respect to these points. These values will be stored in an array and saved in the user's directory in the file ANS. The columns of the array will be:

x	w	dw/dx	M	V

12.5 PROGRAM winkler.m

Program winkler.m calculates deflections, slopes, shears, and moments for beams on an elastic (Winkler) foundation. The following programs and data files are needed in your working directory to run an analysis:

Input Data	User's INCLUDEs	Supplied Functions
MESH	INITIAL.m	SF.m
QUAD	COEF.m	sGAUSS.m

12.5.1 Flow Chart

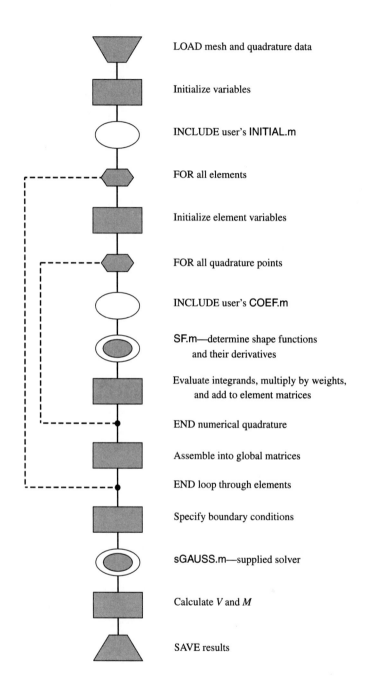

LOAD mesh and quadrature data

Initialize variables

INCLUDE user's **INITIAL.m**

FOR all elements

Initialize element variables

FOR all quadrature points

INCLUDE user's **COEF.m**

SF.m—determine shape functions
    and their derivatives

Evaluate integrands, multiply by weights,
    and add to element matrices

END numerical quadrature

Assemble into global matrices

END loop through elements

Specify boundary conditions

sGAUSS.m—supplied solver

Calculate V and M

SAVE results

12.5.2 Code

```
%-------------------
%  program winkler.m
%-------------------
  clear

%---------------
% INPUT DATA
%---------------
  load MESH    -ASCII
  load QUAD    -ASCII

% ---------------
% Define MESH Data
% ---------------
  NUMNP = MESH(1,1);
  NUMEL = NUMNP-1;
  NUMEQ = 2*NUMNP;
  IB    = 4;
  for i=1:NUMNP;
    XORD(i)   = MESH(i+1,1);
    NPcode(i) = MESH(i+1,2);
  end

% ----------------
% Define QUAD Data
% ----------------
  NQPTS = QUAD(1,1);
  for i=1:NQPTS
     GPTS(i)=QUAD(i+1,1);
     GWTS(i)=QUAD(i+1,2);
  end

% ---------------------
% General initialization
% ---------------------
  for I=1:NUMEQ
     LHS(I)=0.0;
     RHS(I)=0.0;
     NPBC(i)   = 0;
     for J=1:4
        SK(I,J)=0.0;
     end
  end

  INITIAL
```

Finite element analysis of a beam on an elastic (Winkler) foundation

$$\frac{d^2}{dx^2}\left(EI\,\frac{d^2w}{dx^2} \right) + kw - q = 0$$

Load data from mesh.m and quadrature data located in working directory.

Variable	Definition
NUMNP	Number of nodal points
NUMEL	Number of elements
NUMEQ	Number of equations
IB	Bandwidth
XORD	Node's x coordinates
NPcode	Nodal point code
GPTS	Gauss point coordinates
GWTS	Gauss point weights

Initialize and set default values.

Variable	Definition
LHS	Left-hand side; w and dw/dx
RHS	Right-hand side, i.e., the loading
NPBC	Nodal point boundary condition
SK	Global stiffness matrix

INCLUDE user's INITIAL.m code for problem initialization.

```
% --------------------------------
% FORMATION OF STIFFNESS MATRIX
% and RIGHT-HAND SIDE
% --------------------------------

  for I=1:NUMEL
% ----------------------------
% Initialize Element Variables
% ----------------------------
    for J=1:4
      for K=1:4
        S1(J,K)=0.0;
        S2(J,K)=0.0;
      end
      Qe(J)=0.0;
    end


% ----------------------------
% Calculate element coordinate
% information
% ----------------------------
    Xa = XORD(I);
    Xb = XORD(I+1);
    RL=Xb-Xa;
    DxDu=RL/2.0;
    DuDx=2.0/RL;


% ----------------------------
% Begin Gaussian Quadrature
% ----------------------------
  for J=1:NQPTS
    u  = GPTS(J);
    Wt = GWTS(J);


%   --------------------
%   Global coordinate of
%   current Gauss point
%   --------------------
    Xg = (Xa+Xb)/2  + (RL/2.0)*u;


%   ----------------------------
%   INCLUDE COEF.m
%   Defines: EIx, Kx, and Qx
%   ----------------------------
    COEF
```

Form [SK] and {RHS}

Begin integration element by element.

Initialize element matrices, where

$$S1(J,K) = \int \{N''\}EI\lfloor N'' \rfloor\, dx$$

$$S2(J,K) = \int \{N\}k\lfloor N \rfloor\, dx$$

$$Qe(J) = \int \{N\}q\, dx$$

Calculate element dimensions and mapping parameters.

  Xa = Global coordinate of left node
  Xb = Global coordinate of right node
  RL = Length of element
DxDu = dx/du
DuDx = du/dx

Begin loop through quadrature points

  u = Gauss coordinate
Wt = Gauss weight

Xg = global coordinate at center of element

INCLUDE user's COEF.m code to define values of EI, k, and q.

```
% ------------------------------
% Calculate shape functions wrt
% x at current Gauss point
% ------------------------------
   No(1) = SF(0,1,u);
   No(2) = SF(0,2,u)*(DxDu);
   No(3) = SF(0,3,u);
   No(4) = SF(0,4,u)*(DxDu);

   Nxx(1) = SF(2,1,u)*(DuDx)^2;
   Nxx(2) = SF(2,2,u)*(DuDx)^2*(DxDu);
   Nxx(3) = SF(2,3,u)*(DuDx)^2;
   Nxx(4) = SF(2,4,u)*(DuDx)^2*(DxDu);




% --------------------------
% Element stiffness matrices
% --------------------------
   Wt = GWTS(J);
    for K=1:4
     for L=1:4
       S1(K,L)=S1(K,L) + ...
       Wt*Nxx(K)*EIx*Nxx(L)*DxDu;
       S2(K,L)=S2(K,L) + ...
       Wt*No(K)*Kx*No(L)*DxDu;
     end
     Qe(K)=Qe(K) + Wt*No(K)*Qx*DxDu;
    end
   end
% --------- Quadrature now complete

% ----------------------------
% Assemble into global matrices
% ----------------------------
   K1=2*I-2;
   for K=1:4
     K1=K1+1;
     L1=0;
     for L=K:4
      L1=L1+1;

      SK(K1,L1)=SK(K1,L1) + ...
               S1(K,L)+S2(K,L);
     end

     RHS(K1) =  RHS(K1) + Qe(K);
    end
   end
% ---- Global Matrices are assembled
```

Convert shape functions to x coordinates.

$$No(1) = N_1(u)$$
$$No(2) = N_2(u)\frac{dx}{du}$$

Similar for No(3) and No(4).

$$Nxx(1) = \frac{d^2}{dx^2}(N_1) = \frac{d^2 N_1}{du^2}\left(\frac{du}{dx}\right)^2$$

$$Nxx(2) = \frac{d^2}{dx^2}\left(N_2\frac{dx}{du}\right)$$

$$= \frac{d^2 N_1}{du^2}\left(\frac{du}{dx}\right)^2\frac{dx}{du}$$

Similar for Nxx(3) and Nxx(4).

Evaluate the integrands at current Gaussian point and add to quadrature sum.

$$S1(J,K) = Wt\left\{\frac{d^2 N}{dx^2}\right\} EI \left\lfloor\frac{d^2 N}{dx^2}\right\rfloor \frac{dx}{du}$$

$$S2(J,K) = Wt\left\{N\right\} k\lfloor N\rfloor \frac{dx}{du}$$

$$Qe(J) = Wt\{N\} q \frac{dx}{du}$$

Assemble in banded form, bandwidth = 4.

K1 = 4 equation numbers: $(2I - 1)$ to $(2I + 2)$
L1 = 4 column numbers: 1–4

```
% -------------------------
% BOUNDARY CONDITIONS
% -------------------------
  B = 1.0E+06;
  for I=1:NUMNP
   if NPBC(I) == 1 | NPBC(I) == 3
     I1 = 2*I-1;
     SK(I1,1)=SK(I1,1)*B;
     RHS(I1)=LHS(I1)*SK(I1,1);
   end
   if NPBC(I) == 2 | NPBC(I) == 3
     I2=2*I;
     SK(I2,1)=SK(I2,1)*B;
     RHS(I2)=LHS(I2)*SK(I2,1);
   end
  end


% -------------------------
% CALL EQUATION SOLVER
% -------------------------
  LHS = sGAUSS(SK,RHS,NUMEQ,IB);


% --------------------------------------
% Nodal values for w and dw/dx are now
% in LHS. Use these values to calculate
% shear and moment at center of element.
%
% First calculate shape functions
% and their derivatives. Ic is
% counter for number elements used.
% --------------------------------------
  Ic = 0;
  for I=1:NUMEL
    Xa = XORD(I);
    Xb = XORD(I+1);
    RL=Xb-Xa;
    Xg=(Xa+Xb)/2;
    DuDx=2.0/RL;
    DxDu=RL/2.0;

    No(1) = SF(0,1,0);
    No(2) = SF(0,2,0)*(DxDu);
    No(3) = SF(0,3,0);
    No(4) = SF(0,4,0)*(DxDu);

    Nx(1) = SF(1,1,0)*DuDx;
    Nx(2) = SF(1,2,0)*(DxDu)*DuDx;
    Nx(3) = SF(1,3,0)*DuDx;
    Nx(4) = SF(1,4,0)*(DxDu)*DuDx;
```

Specify boundary conditions.

NPBC	w	dw/dx	V	M
0	U	U	K	K
1	K	U	U	K
2	U	K	K	U
3	K	K	U	U

where
K = known
U = unknown

Call equation solver for symmetric, banded storage.

Begin calculations of shear and bending moments at center of each element.

Determine element dimensions and mapping parameters:

Ic = Counter for number of entries in table
Xa = x coordinate of left node
Xb = x coordinate of right node
RL = Element's length
Xg = Coordinate at center of element
DuDx = du/dx
DxDu = dx/du

Determine values of shape functions and their derivatives at center of each element.

$$No(I) = \lfloor N \rfloor$$

$$Nx(I) = \lfloor dN/dx \rfloor$$

```
    Nxx(1)      = SF(2,1,0)*DuDx^2;
    Nxx(2)      = SF(2,2,0)*DxDu*DuDx^2;
    Nxx(3)      = SF(2,3,0)*DuDx^2;
    Nxx(4)      = SF(2,4,0)*DxDu*DuDx^2;

    Nxxx(1)     = SF(3,1,0)*DuDx^3;
    Nxxx(2)     = SF(3,2,0)*DxDu*DuDx^3;
    Nxxx(3)     = SF(3,3,0)*DuDx^3;
    Nxxx(4)     = SF(3,4,0)*DxDu*DuDx^3;
% -------------------------------------
% Calculate deflection, slope, shear, and
% moment at center of element.
% -------------------------------------
    COEF
    W=0;
    S=0;
    V=0;
    M=0;
    for J=1:4
        W = W + No(J)*LHS(2*I-2+J);
        S = S + Nx(J)*LHS(2*I-2+J);
        M = M + EIx*Nxx(J)*LHS(2*I-2+J);
        V = V + EIx*Nxxx(J)*LHS(2*I-2+J)...
                +dEIx*Nxx(J)*LHS(2*I-2+J);
    end
% -------------------------
% Save center values except
% for a pin-element.
% -------------------------
    if EIx   > 0
        Ic = Ic + 1;
        ANS(Ic,1)    =   Xg;
        ANS(Ic,2)    =   W;
        ANS(Ic,3)    =   S;
        ANS(Ic,4)    =   M;
        ANS(Ic,5)    =   V;
    end
  end
% -------------------------
% Solution is now complete.
% Save values.
% -------------------------
    save ANS ANS -ASCII

NOTE = ...
['Solution is saved in file:   ';  ...
 '            ANS                ';  ...
 'where the columns are:        ';  ...
 '                              ';  ...
 '[ x | w | dw/dx | M | V ]     ';  ...
 '                              ';  ...
 'To plot, e.g.  w vs x,  type ';  ...
 'plot(ANS(:,1),ANS(:,2))       ']
```

$Nxx(I) = \lfloor d^2/dx^2 \rfloor$

$Nxxx(I) = \lfloor d^3N/dx^3 \rfloor$

Note: $SF(I,J,0) = \dfrac{d^J N_I(u = 0)}{du^J}$

Calculate deflection, slope, moment, and shear at center of element.

Use COEF to obtain EI and dEI/dx.

W = Deflection
S = Slope, dw/dx
M = Moment
V = Shear

Note: $V = (dEI/dx)\lfloor N'' \rfloor \{W\} + EI \lfloor N''' \rfloor \{W\}$

because $\dfrac{d}{dx}\left(EI\dfrac{d^2w}{dx^2}\right) = \dfrac{dEI}{dx}\dfrac{d^2w}{dx^2} + EI\dfrac{d^3w}{dx^3}$

If $EIx = 0$, element is fictitious, used to represent a pin. Do not record values associated with it.

Save values in user's directory under ANS, where columns are

$$x \quad w \quad dw/dx \quad M \quad V$$

12.5.3 Auxiliary Code

SF.m

```
function s  = SF(D,n,u)
%----------------------------------------
%   D   =   derivative
%   n   =   shape function number
%   u   =   Gauss coordinate
%----------------------------------------

  if n == 1  % Shape function 1
    if D == 0
       s = (1/4)*(u+2)*(u-1)^2;
    elseif D == 1
       s = (1/4)*(3*u^2-3);
    elseif D == 2
       s = (1/4)*(6*u);
    elseif D == 3
       s = (1/4)*(6);
    end

  elseif n == 2  % Shape function 2
    if D == 0
       s = (1/4)*(u+1)*(u-1)^2;
    elseif D == 1
       s = (1/4)*(3*u^2-2*u-1);
    elseif D == 2
       s = (1/4)*(6*u-2);
    elseif D == 3
       s = (1/4)*(6);
    end

  elseif n == 3  % Shape function 3
    if D == 0
       s = -(1/4)*(u-2)*(u+1)^2;
    elseif D == 1
       s = -(1/4)*(3*u^2-3);
    elseif D == 2
       s = -(1/4)*(6*u);
    elseif D == 3
       s = -(1/4)*(6);
    end

  elseif n == 4  % Shape function 4
    if D == 0
       s = (1/4)*(u-1)*(u+1)^2;
    elseif D == 1
       s = (1/4)*(3*u^2+2*u-1);
    elseif D == 2
       s = (1/4)*(6*u+2);
    elseif D == 3
       s = (1/4)*(6);
    end
end
```

Shape function evaluations

$$s = \frac{d^D N_n(u)}{du^D}$$

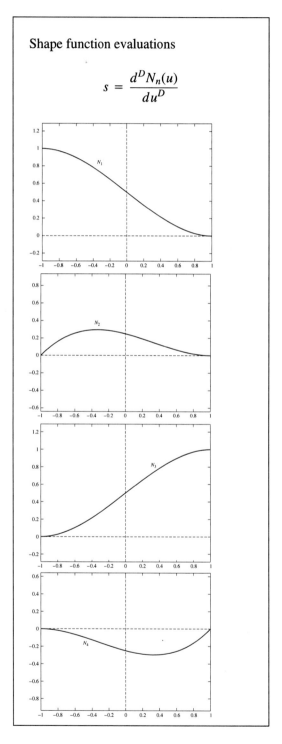

12.5.4 Quadrature Data

QUAD data file

```
%------------------------------------------------
%     Number of points     Dummy number
%------------------------------------------------
             4                    0
%================================================
%     Coordinates           Weights
%------------------------------------------------
     -0.861136311594953     0.347854845137454
     -0.339981043584856     0.652145154862546
      0.861136311594953     0.347854845137454
      0.339981043584856     0.652145154862546
%------------------------------------------------
```

12.5.5 Test Problem. To create one or more problems for which our finite element code will produce the exact solution, it is necessary to decouple the flexural stiffness from the foundation stiffness. We present two such problems.

Cantilever Beam with k = 0

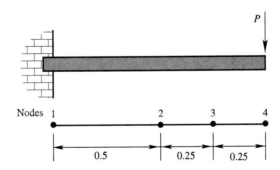

Consider the cantilever beam and its finite element divisions as shown in the figure. From elementary mechanics of solids, we know that the exact solution will be a cubic with a maximum downward deflection at the end equal to

$$y_{max} = -\frac{1}{3}\frac{PL^3}{EI}$$

If we set EI, L, and P equal to unity, our finite element solution should give us $-1/3$ as the deflection at node 4. The data files and user INCLUDE codes for this problem are as follows:

MESH

```
%------------------------
%  MESH  data
%------------------------
%  NUMNP + Dummy
%------------------------
     4      0
%------------------------
%  XORD  NPcode
%------------------------
    0.0     1
    0.5     0
    0.75    0
    1.0     2
```

COEF.m

```
%-------------------------
% Coefficients for
%    winkler.m
%-------------------------
%
%  Set:
%  EIx, Kx, Qx
%
%  dEIx = dEI/dx
%
% Note:
%  Xg = x-coordinate of
%       current point
%-------------------------

     EIx  = 1;
     Kx   = 0;
     dEIx = 0;
     Qx   = 0;
```

INITIAL.m

```
%------------------------------
%          INITIAL.m
%------------------------------
% Must specify nondefault
% values (all zeros) for node n:
%------------------------------
%                  w      dw/dx
%------------------------------
% NPBC(n) = 0      U        U
%         = 1      K        U
%         = 2      U        K
%         = 3      K        K
%------------------------------
% RHS(2n-1) = nodal force
% RHS(2n)   = nodal moment
%------------------------------
% LHS(2N-1) = nodal deflection
% LHS(2N)   = nodal dw/dx
%------------------------------



%------------------------------
% Simply supported beam with
% concentrated unit load at
% midspan (node 2)
%------------------------------
  NPBC(1)      = 1; % Left node
  NPBC(4)      = 1; % Right node
  RHS(2*2-1)   = -1; % Force at
                     % midspan.
```

The solution is saved in the user's working directory as **ANS**. It will contain the following values, which correspond to the exact solution within the round-off error of the numerical calculations. Note that these values correspond to points at the center of the elements.

ANS

XORD	W	dW/dx	M	V
2.500e-01	-2.8645875e-02	-2.1875013e-01	-7.500e-01	1.000e-00
6.250e-01	-1.5462248e-01	-4.2968763e-01	-3.750e-01	1.000e-00
8.750e-01	-2.7115897e-01	-4.9218763e-01	-1.250e-01	1.000e-00

Foundation Support Only

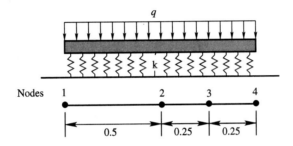

We now consider a beam resting on an elastic foundation with no other support. It has a uniform load as shown; hence, it will deform without bending and have a uniform deflection equal to $w = q/k$. For $q = -1$, $k = 1$, and $EI = 1$, the exact solution for the deflection will be -1.

The following data files and user INCLUDE codes are needed for this problem:

MESH

```
%--------------------------
% MESH  data
%--------------------------
% NUMNP + Dummy
%--------------------------
      4      0
%--------------------------
% XORD   NPcode
%--------------------------
     0.0      1
     0.5      0
     0.75     0
     1.0      2
```

COEF.m

```
%----------------------
% Coefficients for
%    winkler.m
%----------------------
%
% Set:
% EIx, Kx, Qx
%
% dEIx =  dEI/dx
%
% Note:
% Xg = x-coordinate of
%      current point.
%----------------------

        EIx  = 1;
        Kx   = 1;
        dEIx = 0;
        Qx   = 1;
```

INITIAL.m

```
%---------------------------------
%            INITIAL.m
%---------------------------------
% Must specify nondefault
% values (all zeros) for node n:
%---------------------------------
%                    w       dw/dx
%---------------------------------
% NPBC(n) = 0       U        U
%         = 1       K        U
%         = 2       U        K
%         = 3       K        K
%---------------------------------
% RHS(2n-1) = nodal force
% RHS(2n)   = nodal moment
%---------------------------------
% LHS(2N-1) = nodal deflection
% LHS(2N)   = nodal dw/dx
%---------------------------------




%---------------------------------
% Simply supported beam with
% concentrated unit load at
% mid span (node 2)
%---------------------------------
   NPBC(1)       = 0; % Left node
   NPBC(4)       = 0; % Right node
```

Note: The NPBC values initialized here are the default values. They did not need to be specified, but are for emphasis.

The solution is saved in the user's working directory as **ANS**. It will contain the following values, which correspond to the exact solution within the round-off error of the numerical calculations. Note that these values correspond to points at the center of the elements.

ANS

2.500e-01	-1.00e+00	-2.725021e-13	-1.406486e-15	1.223272e-14
6.250e-01	-1.00e+00	-2.714533e-13	4.536202e-15	-1.273049e-14
8.750e-01	-1.00e+00	-2.709269e-13	5.787325e-15	1.247846e-14
XORD	W	dW/dx	M	V

EXERCISES

Study Problems

S1. From basic principles of work and energy, derive Eq. 12.3 for the potential energy of the beam on an elastic foundation.

S2. Begin with Eq. 12.3 and determine the governing equation and boundary conditions if the potential energy is to have a stationary value.

S3. Derive each of the four shape functions given in Eq. 12.7.

S4. Consider the transformation of coordinates given in Eq. 12.21 and the shape functions given in Eq. 12.7. Based on this, would it be proper to say we are using a one-dimensional isoparametric element? Explain your answer.

S5. Verify all equations appearing in Eq. 12.22.

Numerical Experiments and Code Development

N1. Change the test problem for the cantilever beam to be
   (a) Built-in at left end and roller at right end. Load in center.
   (b) Built-in at both ends. Load in center.

N2. Simplify winkler.m by noting that $(du/dx)(dx/du) = 1$.

N3. Write a MATLAB code to read ANS, and then plot either w, dw/dx, M, or V depending on what the user specifies, or plot all of the above in sequence.

N4. Add to winkler.m one more column in ANS that gives the force per unit length along the beam created by the foundation.

Project

P1. For each beam shown in the figure, plot the deflection, and the moment and shear diagrams. Also determine the maximum and minimum values of w, M, and V and the locations where they occur. For all beams, set $L = 1$. The other nondimensional parameters are as follows:

Beam	EI	k	q	P	M
a	1	20,000	1	0	0
b	1	1,000	1	0	0
c	1	1,000	1	0	0
d	1	20,000	1	0	0
e	1	20,000	1	0	0
f	1	10,000	0	1	0.1PL
g	$1 - (x^3)/2$	1,000	$-1 + x/2$	0	0
h	1	$100,000*(x^4)$	0	−1	0

The loads, q, P, and M have the directions shown on each beam. Report your answers in terms of representative values of L, EI, and q or P. For beam g, use symmetry. On other problems with symmetry, the choice is yours to make.

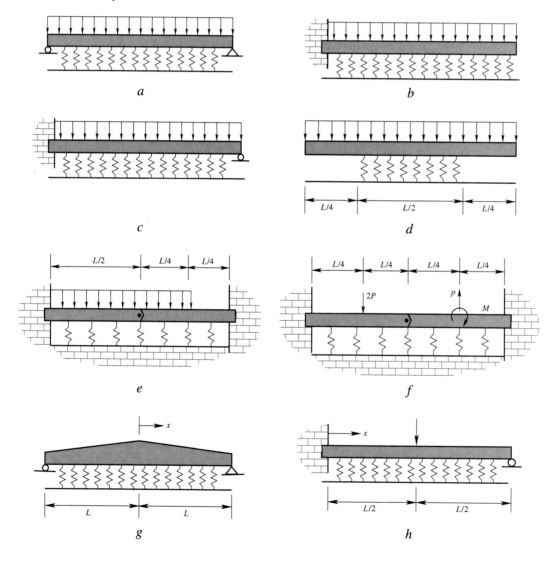

a

b

c

d

e

f

g

h

EQUATION SOLVERS AND COMPACT STORAGE

It is the nature of finite element approximations that any two nodes are mathematically connected only when they share an element. This gives rise to global stiffness matrices that are always sparse (having many more zero terms than nonzero terms) and often banded (all nonzero terms appear within a certain distance of the diagonal of the matrix). For this reason, equation solvers are written to take advantage of this sparseness to save computer storage and computer time. Because of the large number of unknowns in most finite element analyses, a great deal of research has been conducted to determine efficient solution algorithms for these equations. These algorithms fall into two general classes: direct solvers and indirect solvers. Gaussian elimination is the prototype of most direct solvers. The indirect solvers make use of iterative methods and include the Gauss-Seidel method and the conjugate gradient method. For very large systems of equations, the indirect solvers usually prove more efficient. However, for the size of problems considered in this text, Gaussian elimination is the most practical approach to use.

A.1 GAUSSIAN ELIMINATION

There are many direct methods, all of which have special advantages under certain conditions. However, what we call simple Gaussian elimination will suffice for our purposes. We describe it using the following example.

Consider the matrix equation

$$
\begin{bmatrix}
4 & -4 & -4 & 0 & 0 & 0 & 0 \\
-4 & 7 & -2 & -3 & 0 & 0 & 0 \\
-4 & -2 & 19 & -3 & -6 & 0 & 0 \\
0 & -3 & -3 & 35 & 8 & -15 & 0 \\
0 & 0 & -6 & 8 & 34 & 28 & -2 \\
0 & 0 & 0 & -15 & 28 & 51 & -6 \\
0 & 0 & 0 & 0 & -2 & -6 & 21
\end{bmatrix}
\begin{Bmatrix}
X_1 \\ X_2 \\ X_3 \\ X_4 \\ X_5 \\ X_6 \\ X_7
\end{Bmatrix}
=
\begin{Bmatrix}
-16 \\ -8 \\ 19 \\ 119 \\ 170 \\ 120 \\ 3
\end{Bmatrix}
$$

We now replace the second equation with the sum of it and the first equation, and replace the third equation with the sum of it and the first equation. This gives us

$$
[A] = \begin{bmatrix}
4 & -4 & -4 & 0 & 0 & 0 & 0 \\
0 & 3 & -6 & -3 & 0 & 0 & 0 \\
0 & -6 & 15 & -3 & -6 & 0 & 0 \\
0 & -3 & -3 & 35 & 8 & -15 & 0 \\
0 & 0 & -6 & 8 & 34 & 28 & -2 \\
0 & 0 & 0 & -15 & 28 & 51 & -6 \\
0 & 0 & 0 & 0 & -2 & -6 & 21
\end{bmatrix}
\quad
\{B\} = \begin{Bmatrix}
-16 \\
-24 \\
3 \\
119 \\
170 \\
120 \\
3
\end{Bmatrix}
$$

Notice that we have succeeded in making all terms beneath the first diagonal equal to zero. This was the purpose of our additions. We now do the same for the second diagonal. This time we must multiply the second row by 2 before we add it to the third row to obtain a zero beneath the diagonal. We can, however, simply add the second equation to the fourth equation to obtain our desired zero term. By doing this, we obtain

$$
[A] = \begin{bmatrix}
4 & -4 & -4 & 0 & 0 & 0 & 0 \\
0 & 3 & -6 & -3 & 0 & 0 & 0 \\
0 & 0 & 3 & -9 & -6 & 0 & 0 \\
0 & 0 & -9 & 32 & 8 & -15 & 0 \\
0 & 0 & -6 & 8 & 34 & 28 & -2 \\
0 & 0 & 0 & -15 & 28 & 51 & -6 \\
0 & 0 & 0 & 0 & -2 & -6 & 21
\end{bmatrix}
\quad
\{B\} = \begin{Bmatrix}
-16 \\
-24 \\
-45 \\
95 \\
170 \\
120 \\
3
\end{Bmatrix}
$$

Next we follow the same procedure to obtain zeros beneath the remaining diagonals as shown in the following sequence of matrices. Note that for emphasis, we have left a blank space rather than a zero for each of the eliminated terms.

$$[A] = \begin{bmatrix} 4 & -4 & -4 & 0 & 0 & 0 & 0 \\ & 3 & -6 & -3 & 0 & 0 & 0 \\ & & 3 & -9 & -6 & 0 & 0 \\ & & & 5 & -10 & -15 & 0 \\ & & & -10 & 22 & 28 & -2 \\ & & & -15 & 28 & 51 & -6 \\ & & & 0 & -2 & -6 & 21 \end{bmatrix} \qquad \{B\} = \begin{Bmatrix} -16 \\ -24 \\ -45 \\ -40 \\ 0 \\ 0 \\ 3 \end{Bmatrix}$$

$$[A] = \begin{bmatrix} 4 & -4 & -4 & 0 & 0 & 0 & 0 \\ & 3 & -6 & -3 & 0 & 0 & 0 \\ & & 3 & -9 & -6 & 0 & 0 \\ & & & 5 & -10 & -15 & 0 \\ & & & & 2 & -2 & -2 \\ & & & & -2 & 6 & -6 \\ & & & & -2 & -6 & 21 \end{bmatrix} \qquad \{B\} = \begin{Bmatrix} -16 \\ -24 \\ -45 \\ -40 \\ 0 \\ 0 \\ 3 \end{Bmatrix}$$

$$[A] = \begin{bmatrix} 4 & -4 & -4 & 0 & 0 & 0 & 0 \\ & 3 & -6 & -3 & 0 & 0 & 0 \\ & & 3 & -9 & -6 & 0 & 0 \\ & & & 5 & -10 & -15 & 0 \\ & & & & 2 & -2 & -2 \\ & & & & & 4 & -8 \\ & & & & & & -8 & 19 \end{bmatrix} \qquad \{B\} = \begin{Bmatrix} -16 \\ -24 \\ -45 \\ -40 \\ 0 \\ 0 \\ 3 \end{Bmatrix}$$

$$[A] = \begin{bmatrix} 4 & -4 & -4 & 0 & 0 & 0 & 0 \\ & 3 & -6 & -3 & 0 & 0 & 0 \\ & & 3 & -9 & -6 & 0 & 0 \\ & & & 5 & -10 & -15 & 0 \\ & & & & 2 & -2 & -2 \\ & & & & & 4 & -8 \\ & & & & & & 3 \end{bmatrix} \qquad \{B\} = \begin{Bmatrix} -16 \\ -24 \\ -45 \\ -40 \\ 0 \\ 0 \\ 3 \end{Bmatrix}$$

Our final equation can now be written as

$$
\begin{bmatrix}
4 & -4 & -4 & 0 & 0 & 0 & 0 \\
 & 3 & -6 & -3 & 0 & 0 & 0 \\
 & & 3 & -9 & -6 & 0 & 0 \\
 & & & 5 & -10 & -15 & 0 \\
 & & & & 2 & -2 & -2 \\
 & & & & & 4 & -8 \\
 & & & & & & 3
\end{bmatrix}
\begin{Bmatrix}
X_1 \\ X_2 \\ X_3 \\ X_4 \\ X_5 \\ X_6 \\ X_7
\end{Bmatrix}
=
\begin{Bmatrix}
-16 \\ -24 \\ -45 \\ -40 \\ 0 \\ 0 \\ 3
\end{Bmatrix}
$$

We now note that the last equation represents an equation with one unknown, i.e.,

$$3(X_7) = 3$$

Hence, $X_7 = 1$. With this now known, the next to last equation has only one unknown, i.e.,

$$4(X_6) - 8(X_7) = 0$$

Hence, $X_6 = 2$. We continue in this manner, moving up the list of equations until we reach the first equation, with X_1 the only unknown. The final solution is found to be

$$\left\{ \begin{array}{c} X_1 \\ \hline X_2 \\ \hline X_3 \\ \hline X_4 \\ \hline X_5 \\ \hline X_6 \\ \hline X_7 \end{array} \right\} = \left\{ \begin{array}{c} 1 \\ \hline 2 \\ \hline 3 \\ \hline 4 \\ \hline 3 \\ \hline 2 \\ \hline 1 \end{array} \right\}$$

The process of creating zeros under all diagonal terms, starting with the first and moving down the diagonal, is referred to as *forward elimination*. The process of solving for the unknowns, starting with the last equation and moving up the set of equations, is referred to as *back substitution*.

A.2 BANDED STORAGE

We now note that all terms in any row of [A] beyond two locations on either side of the diagonal are zero. Furthermore, these terms remained zero during the Gaussian elimination. That is, we were able to solve our equation using only the following terms in our matrix:

Original matrix

4	−4	−4				
−4	7	−2	−3			
−4	−2	19	−3	−6		
	−3	−3	35	8	−15	
		−6	8	34	28	−2
			−15	28	51	−6
				−2	−6	21

Original matrix

Final matrix

4	−4	−4				
0	3	−6	−3			
0	0	3	−9	−6		
	0	0	5	−10	−15	
		0	0	2	−2	−2
			0	0	4	−8
				0	0	3

Final matrix

This is true for any banded matrix. Thus, it is necessary only to store the terms that are inside the band of nonzero terms on each side of the diagonal. How this is accomplished can be visualized as sliding each of the rows so that all the diagonal terms appear in the same column. Thus,

D

·	·	4	−4	−4
·	−4	7	−2	−3
−4	−2	19	−3	−6
−3	−3	35	8	−15
−6	8	34	28	−2
−15	28	51	−6	·
−2	−6	21	·	·

Original matrix

D

·	·	4	−4	−4
·	0	3	−6	−3
0	0	3	−9	−6
0	0	5	−10	−15
0	0	2	−2	−2
0	0	4	−8	·
0	0	3	·	·

Final matrix

Here we have shown the calculated zeros and used [·] to indicate locations not in the original matrix but that must be included in the compact matrix. The diagonal terms appear in the center column (marked with a D). This new, compact matrix has the same number of rows as the original matrix, but only as many columns as the width of the band of nonzero terms along the diagonal—the "bandwidth" of the matrix. This savings in storage can become significant. Consider, for example, a matrix of 2000 equations with a bandwidth of 30!

We now consider an additional savings in storage that is often possible. The compact storage arrangement shown is sufficient and necessary for nonsymmetric matrices; however, the preceding matrix happens to be symmetric. We can therefore make the storage even more compact. This is possible because, during a Gaussian elimination, the symmetry of the matrix is not destroyed until the lower terms are replaced with zeros (observe the sequence shown earlier). Thus, during the elimination process, any term beneath the diagonal is known to be either zero or equal to its symmetric counterpart above the diagonal. Knowing this, we can store the preceding symmetric matrix as follows:

D		
4	−4	−4
7	−2	−3
19	−3	−6
35	8	−15
34	28	−2
51	−6	·
21	·	·

Original matrix

D		
4	−4	−4
3	−6	−3
3	−9	−6
5	−10	−15
2	−2	−2
4	−8	·
3	·	·

Final matrix

For this storage, the diagonal terms appear in the first column and all terms in the other columns represent the upper half of the matrix. For this storage, the bandwidth is defined as 3.

Before presenting the codes for Gaussian elimination, we make two observations. The first is that most codes for Gaussian elimination conduct row and/or column interchanges before the elimination process begins. This is done to minimize the chance that a zero will turn up on the diagonal during the elimination. For finite element analyses, at least those considered in this text, that is not a concern and will not be covered.

The second observation is that the following codes are written in MATLAB script. This is in keeping with the way all codes have been written in this text so that the reader can understand the numerical algorithm and can convert the codes easily to other languages. However, these codes are rather slow when used with MATLAB, and the user might want to convert the FEM programs in order to make use of MATLAB's own equation solvers.

A.3 CODES

We present three codes for the solution of $[A]\{Y\} = \{F\}$ by Gaussian elimination. The first is the prototype code, GAUSS.m, where $[A]$ is stored as a square matrix. The other two codes are nGAUSS.m, where $[A]$ is stored as a banded, nonsymmetric matrix, and sGAUSS.m, where $[A]$ is stored as a banded symmetric matrix. Program GAUSS.m is not used in any of the codes in the text but is presented as the foundation for the other two codes. The following flow chart describes all three codes.

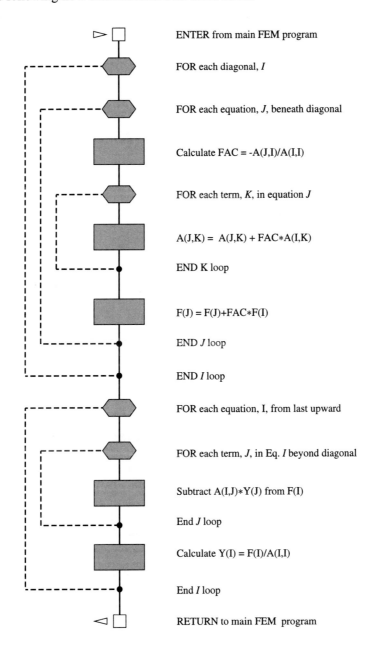

ENTER from main FEM program

FOR each diagonal, *I*

FOR each equation, *J*, beneath diagonal

Calculate FAC = -A(J,I)/A(I,I)

FOR each term, *K*, in equation *J*

A(J,K) = A(J,K) + FAC*A(I,K)

END K loop

F(J) = F(J)+FAC*F(I)

END *J* loop

END *I* loop

FOR each equation, I, from last upward

FOR each term, *J*, in Eq. *I* beyond diagonal

Subtract A(I,J)*Y(J) from F(I)

End *J* loop

Calculate Y(I) = F(I)/A(I,I)

End *I* loop

RETURN to main FEM program

Gauss elimination for full storage

```
%-----------------------------------
  function  Y = GAUSS(A,F,neq)
%-----------------------------------

%-----------------------------
% Begin forward elimination
%-----------------------------
  for I = 1:(neq-1)
     for J=(I+1):neq
        FAC=-A(J,I)/A(I,I);
        for K=I:neq
           A(J,K)=A(J,K)+FAC*A(I,K);
        end
        F(J)=F(J)+FAC*F(I);
     end
  end

%-----------------------------
% Begin back substitution
%-----------------------------
  Y(neq)=F(neq)/A(neq,neq);
     for Iback=2:neq
        I=neq-Iback+1;
        for J=(I+1):neq
           F(I)=F(I)-A(I,J)*Y(J);
        end
        Y(I)=F(I)/A(I,I);
     end
```

The next two codes have exactly the same pattern as the preceding code. However, when an element within the [A] matrix is needed, its column number must correspond to that used in the banded form. This number is determined relative to where the diagonal column appears in these two banded forms, which is designated in both as Idiag. For symmetric storage, this will be the first column, and for nonsymmetric storage it will be the middle column. With this designation, the term A(I,K) will appear at location A(I,Kc), where Kc = Idiag-(K-I). Note that the numerical value of Kc depends on the particular row being addressed. When two different rows are involved, the columns are named IKc and JKc for rows I and J, respectively. The lowercase c stands for "compact." The only additional change is that the beginning and end of the J loops must be compatible with the bandwidth and the compact storage used; hence, we calculate and use Jbgn and Jend for this purpose.

For banded,nonsymmetric storage

```
%----------------------------------
   function  Y = nGAUSS(A,F,neq,IB)
%----------------------------------
%   Solves [A]{Y} = {F}
%   for banded nonsymmetric matices
%----------------------------------

   Idiag=(IB-1)/2+1;
%----------------------
% Begin forward elimination
%----------------------
   for I = 1:(neq-1);

     Jend=neq;
     if Jend > I+(Idiag-1)
        Jend = I+(Idiag-1);
     end

     for J=(I+1):Jend
       Kc=Idiag-(J-I);

       FAC=-A(J,Kc)/A(I,Idiag);

       Kbgn=I;
       Kend=Jend;
       for K=Kbgn:Kend
         IKc=Idiag+(K-I);
         JKc=Idiag+(K-J);
         A(J,JKc)=A(J,JKc)+FAC*A(I,IKc);
       end
       F(J)=F(J)+FAC*F(I);
     end
   end

%----------------------
% Begin back substitution
%----------------------
   Y(neq)=F(neq)/A(neq,Idiag);
   for Iback=2:neq
     I=neq-Iback+1;
     Jend=neq;
     if Jend > I+(Idiag-1)
        Jend = I+(Idiag-1);
     end
     for J=(I+1):Jend
       Jc = Idiag+(J-I);
       F(I)=F(I)-A(I,Jc)*Y(J);
     end
     Y(I)=F(I)/A(I,Idiag);
   end
```

For banded, symmetric storage

```
%------------------------------
   function  Y = sGAUSS(A,F,neq,IB)
%------------------------------
%   Solves [A]{Y} = {F}
%   for banded symmetric matrices
%------------------------------

   Idiag=1;
%----------------------
% Begin forward elimination
%----------------------
   for I = 1:(neq-1)

     Jend=neq;
     if Jend > I+(IB-1)
        Jend = I+(IB-1);
     end

     for J=(I+1):Jend
       Kc=J-I+1;
       Aji=A(I,Kc);
       FAC=-Aji/A(I,Idiag);

       Kbgn=J;
       Kend=Jend;
       for K=Kbgn:Kend
         IKc=K-I+1;
         JKc=K-J+1;
         A(J,JKc)=A(J,JKc)+FAC*A(I,IKc);
       end
       F(J)=F(J)+FAC*F(I);
     end
   end

%----------------------
% Begin back substitution
%----------------------
   Y(neq)=F(neq)/A(neq,Idiag);
   for Iback=2:neq
     I=neq-Iback+1;
     Jend=neq;
     if Jend > I+IB-1
        Jend = I+IB-1;
     end
     for J=(I+1):Jend
       Jc = J-I+1;
       F(I)=F(I)-A(I,Jc)*Y(J);
     end
     Y(I)=F(I)/A(I,Idiag);
   end
```

THE SHAPE FUNCTION ARRAY

When Gaussian quadrature is used for the integration of our finite element equations, integrands must be evaluated at each quadrature point in each element. Thus, the number of numerical calculations can be large. Fortunately, many of the terms in these integrands vary from one quadrature point to another, but not from one element to another. That is, their value for a particular quadrature point is the same for all elements. These terms are the numerical values of each shape function and their derivatives with respect to the coordinates used in the parent element. That is,

$$\lfloor N \rfloor \qquad \left\lfloor \frac{\partial N}{\partial u} \right\rfloor \qquad \left\lfloor \frac{\partial N}{\partial v} \right\rfloor$$

These terms are used to evaluate

$$x = \lfloor N \rfloor \{x\}$$

$$y = \lfloor N \rfloor \{y\}$$

$$p = \lfloor N \rfloor \{p\}$$

$$\frac{\partial x}{\partial u} = \left\lfloor \frac{\partial N}{\partial u} \right\rfloor \left\{ x \right\}$$

$$\frac{\partial x}{\partial v} = \left\lfloor \frac{\partial N}{\partial v} \right\rfloor \left\{ x \right\}$$

$$\frac{\partial y}{\partial u} = \left\lfloor \frac{\partial N}{\partial u} \right\rfloor \left\{ y \right\}$$

$$\frac{\partial y}{\partial v} = \left\lfloor \frac{\partial N}{\partial v} \right\rfloor \left\{ y \right\}$$

where p represents any particular parameter that might be given in terms of nodal point values.

Each of the components of the three row vectors $\lfloor N \rfloor$, $\lfloor \partial N/\partial u \rfloor$, and $\lfloor \partial N/\partial v \rfloor$ must be evaluated at each quadrature point using the shape functions defined for the particular element being employed. However, once evaluated for one element, they can be saved and used repeatedly for all other elements. The numbers are saved in a three-dimensional array that can be visualized as a box, or cabinet, as shown here:

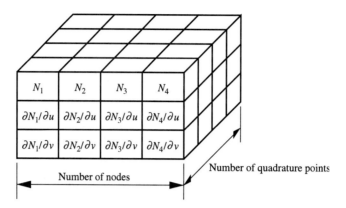

Number of quadrature points

Number of nodes

In this figure, we have assumed a four-node element and four quadrature points. Each of the interior boxes contains a single number representing the quantity shown at a particular quadrature point. As an example of the use of this array, assume your code is performing an element integration and is at quadrature point 3. Picture sliding "file 3" out of the cabinet to create a two-dimensional array as shown

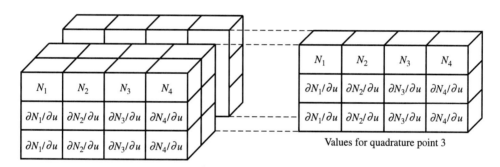

Values for quadrature point 3

This two-dimensional array can now be used to determine the values of all terms that appear in the integrands of the finite element equations. For example,

$$
\begin{Bmatrix} x \\ \dfrac{\partial x}{\partial u} \\ \dfrac{\partial x}{\partial v} \end{Bmatrix}
=
\begin{bmatrix}
N_1 & N_2 & N_3 & N_4 \\
\dfrac{\partial N_1}{\partial u} & \dfrac{\partial N_2}{\partial u} & \dfrac{\partial N_3}{\partial u} & \dfrac{\partial N_4}{\partial u} \\
\dfrac{\partial N_1}{\partial v} & \dfrac{\partial N_2}{\partial v} & \dfrac{\partial N_3}{\partial v} & \dfrac{\partial N_4}{\partial v}
\end{bmatrix}
\begin{Bmatrix} X_1 \\ X_2 \\ X_3 \\ X_4 \end{Bmatrix}
$$

Similarly, the nodal values of the y coordinates could be used to obtain the same three values associated with the y coordinates.

The cabinet can be expanded to include higher-order elements as well as the values used for surface quadrature. Thus, for a six-node quadratic element, the SF array would be

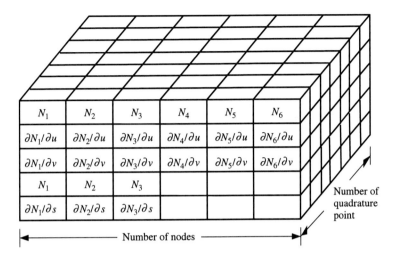

where seven quadrature points for the volume (area) quadrature are assumed. The last two "layers" in the box contain the values for surface (line) quadrature. It is assumed that the number of quadrature points used for these integrations is no more than 7, but could be less. We name this cabinet of numbers the *SF array*. Rather than using the same three-dimensional array for storing both the volume and the surface quadrature data, separate arrays could be used (e.g. SF_v and SF_s).

The *SF* array is calculated in function **SFquad.m**, which is called at the beginning of a finite element code. This function is given the element type by designating the number of nodes per element, NNPE. The number of quadrature points, their coordinates, and their weights are designated within the function. The user can modify this information if desired, such as changing the number of quadrature points for a given element or adding new elements. The function returns to the calling program the arrays SF(I,J,K), WT(I), NUMQPT(I), and NPSIDE(I). These arrays will be defined in the "Notation" section.

B.1 FLOW CHART FOR SFquad.m

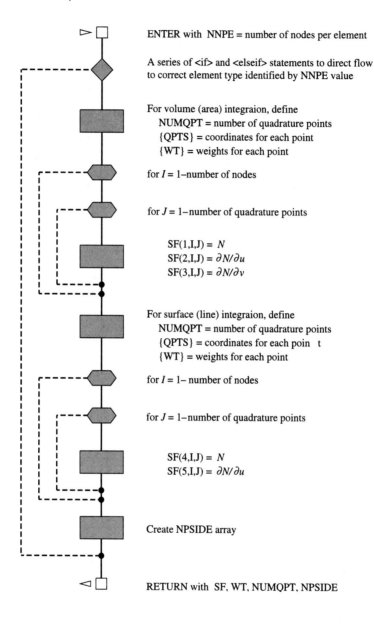

ENTER with NNPE = number of nodes per element

A series of <if> and <elseif> statements to direct flow
to correct element type identified by NNPE value

For volume (area) integraion, define
   NUMQPT = number of quadrature points
   {QPTS} = coordinates for each point
   {WT} = weights for each point

for I = 1–number of nodes

for J = 1–number of quadrature points

$SF(1,I,J) = N$
$SF(2,I,J) = \partial N/\partial u$
$SF(3,I,J) = \partial N/\partial v$

For surface (line) integraion, define
   NUMQPT = number of quadrature points
   {QPTS} = coordinates for each poin t
   {WT} = weights for each point

for I = 1– number of nodes

for J = 1–number of quadrature points

$SF(4,I,J) = N$
$SF(5,I,J) = \partial N/\partial u$

Create NPSIDE array

RETURN with SF, WT, NUMQPT, NPSIDE

B.2 NOTATION
Input and Output Variables

NNPE	Number of nodal points per element
NPSIDE(I,J)	Element numbering for the *J*th node on side *I*
NUMQPT(1)	Number of quadrature points used for volume (or area) integration
NUMQPT(2)	Number of quadrature points used for surface (or line) integration
SF(1,I,J)	Numerical value of element shape function, N, for node *I* at quadrature point *J*
SF(2,I,J)	Numerical value of $\partial N/\partial u$ for node *I* at quadrature point *J*
SF(3,I,J)	Numerical value of $\partial N/\partial v$ for node *I* at quadrature point *J*
SF(4,I,J)	Numerical value of surface shape function, N_s, for node *I* at quadrature point *J*
SF(5,I,J)	Numerical value of $\partial N_s/\partial u$ for node *I* at quadrature point *J*
WT(1,I)	Weight for quadrature point *I*, volume (or area) integration
WT(2,I)	Weight for quadrature point *I*, surface (or line) integration

Selected Temporary Variables

NNPS	Number of nodal points per element side
NSPE	Number of sides per element
QPT(1,I)	Quadrature coordinate of *u* for volume (or area) integration; *or* Quadrature coordinate of *s* for surface (or line) integration
QPT(2,I)	Quadrature coordinate of *v* for volume (or area) integration

Internal Functions

SFN(u,v,J,NNPE)	Numerical value of shape function N for node *J* at point (u, v) for an element with NNPE nodes
SFNu(u,v,J,NNPE)	Numerical value of $\partial N/\partial u$ for node *J* at point (u, v) for an element with NNPE nodes
SFNv(u,v,J,NNPE)	Numerical value of $\partial N/\partial v$ for node *J* at point (u, v) for an element with NNPE nodes
SFL(u,J,NNPS)	Numerical value of shape function N_s for node *J* at point *u* for a line element with NNPS nodes
SFLu(u,J,NNPS)	Numerical value of $\partial N_s/\partial u$ for node *J* at point *u* for a line element with NNPS nodes

B.3 CODE

```
function [SF,WT,NUMQPT,NPSIDE]...
                    = SFquad(NNPE)
%------------------------------

%-----------------
 if NNPE == 3
%-----------------
   NNPS=2;
   NSPE=3;

%----------------------
% SF FOR AREA QUADRATURE
%----------------------
   NUMQPT(1)=1;
   QPT(1,1)=1.0/3.0;
   QPT(1,2)=1.0/3.0;
   WT(1,1) =0.5;

   for I=1:NNPE
    JEND=NUMQPT(1);
    for J=1:JEND
     SF(1,I,J)=...
         SFN(QPT(J,1),QPT(J,2),I,NNPE);
     SF(2,I,J)=...
         SFNu(QPT(J,1),QPT(J,2),I,NNPE);
     SF(3,I,J)=...
         SFNv(QPT(J,1),QPT(J,2),I,NNPE);
    end
   end


%----------------------
% SF for line quadrature
%----------------------
   NUMQPT(2)=1;
   QPT(1,1)= 0;
   WT(2,1) = 2;

   JEND=NUMQPT(2);
   for J=1:JEND
    for I=1:NNPS
     I1=I;
     SF(4,I,J)=SFL (QPT(1,J),I1,NNPS);
     SF(5,I,J)=SFLu(QPT(1,J),I1,NNPS);
    end
   end
```

Function SFquad.m

Input: NNPE
Output: SF, WT, NUMQPT, NPSIDE

Element is a three-node triangle.

Number of nodes per side   = 2
Number of sides per element = 3

Area quadrature:

Specify
(1) Number of quadrature points = 1
(2) u and v coordinates
(3) Weights

Fill in SF matrices.

$$SF(1,I,J) = N_I(u_J, v_J)$$

$$SF(2,I,J) = \frac{\partial N_I}{\partial u}(u_J, v_J)$$

$$SF(3,I,J) = \frac{\partial N_I}{\partial v}(u_J, v_J)$$

Line quadrature:

Specify
(1) Number of quadrature points = 1
(2) u coordinates
(3) Weights

Fill in SF matrices.

$$SF(4,I,J) = NS_I(u_J)$$

$$SF(5,I,J) = \frac{\partial NS_I}{\partial u}(u_J)$$

```
%-------------------
% Define NPSIDE array
%-------------------
  NPSIDE(1,1)=1;
  NPSIDE(1,2)=2;
  NPSIDE(2,1)=2;
  NPSIDE(2,2)=3;
  NPSIDE(3,1)=3;
  NPSIDE(3,2)=1;



%-------------------
 elseif NNPE == 4
%-------------------
 NNPS=2;
 NSPE=4;

%-------------------------
% SF for volume quadrature
%-------------------------
 NUMQPT(1)=4;
 A1=0.5773502692;
 QPT(1,1)=-A1;
 QPT(1,2)=-A1;
 QPT(2,1)=+A1;
 QPT(2,2)=-A1;
 QPT(3,1)=+A1;
 QPT(3,2)=+A1;
 QPT(4,1)=-A1;
 QPT(4,2)=+A1;

 WT(1,1)=1.00;
 WT(1,2)=1.00;
 WT(1,3)=1.00;
 WT(1,4)=1.00;

 JEND=NUMQPT(1);
 for J=1:JEND
   for I=1:4
     SF(1,I,J)=...
       SFN(QPT(J,1),QPT(J,2),I,NNPE);
     SF(2,I,J)=...
       SFNu(QPT(J,1),QPT(J,2),I,NNPE);
     SF(3,I,J)=...
       SFNv(QPT(J,1),QPT(J,2),I,NNPE);
   end
 end
```

Create NPSIDE array that lists element node numbers on each side of element.

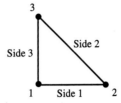

Element is a four-node triangle.

Number of nodes per side   = 2
Number of sides per element = 4

Area quadrature:

Specify
(1) Number of quadrature points = 4

(2) u and v coordinates

(3) Weights

Fill in *SF* matrices.

$$SF(1, I, J) = N_I(u_J, v_J)$$

$$SF(2, I, J) = \frac{\partial N_I}{\partial u}(u_J, v_J)$$

$$SF(3, I, J) = \frac{\partial N_I}{\partial v}(u_J, v_J)$$

```
%-------------------------
% SF for line quadrature
%-------------------------

 NUMQPT(2)=1;
 WT(2,1)=2.0;

 SF(4,1,1)= 0.5;
 SF(4,2,1)= 0.5;
 SF(5,1,1)=-0.5;
 SF(5,2,1)= 0.5;




% -------------------
% Define NPSIDE array
% -------------------
  NPSIDE(1,1)=1;
  NPSIDE(1,2)=2;
  NPSIDE(2,1)=2;
  NPSIDE(2,2)=3;
  NPSIDE(3,1)=3;
  NPSIDE(3,2)=4;
  NPSIDE(4,1)=4;
  NPSIDE(4,2)=1;




%-----------------
  elseif NNPE == 6
%-----------------
  NNPS=3;
  NSPE=3;


% -----------------------
% SF FOR AREA QUADRATURE
% -----------------------
  NUMQPT(1)=7;
  A1=0.059715871789770;
  B1=0.470142064105115;
  A2=0.797426985353087;
  B2=0.101286507323456;
```

Line quadrature:

Number of quadrature points = 1
Weight = 2.0

$SF(4,1,1) = N_1(u = 0)$
$SF(4,2,1) = N_2(u = 0)$
$SF(5,1,1) = dN_1/du(u = 0)$
$SF(5,2,1) = dN_2/du(u = 0)$

Create NPSIDE array for four-node element.

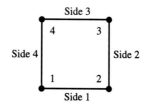

Element is a six-node triangle.

Number of nodes per side   = 3
Number of sides per element = 3

Area quadrature:

Number of quadrature points = 7

Define quadrature coordinates.

```
QPT(1,1)=1.0/3.0;
QPT(1,2)=1.0/3.0;
QPT(2,1)=A1;
QPT(2,2)=B1;
QPT(3,1)=B1;
QPT(3,2)=A1;
QPT(4,1)=B1;
QPT(4,2)=B1;
QPT(5,1)=B2;
QPT(5,2)=A2;
QPT(6,1)=B2;
QPT(6,2)=B2;
QPT(7,1)=A2;
QPT(7,2)=B2;


WT(1,1)=0.1125;
WT(1,2)=0.066197076394253;
WT(1,3)=WT(1,2);
WT(1,4)=WT(1,2);
WT(1,5)=0.062969590272413;
WT(1,6)=WT(1,5);
WT(1,7)=WT(1,5);


JEND=NUMQPT(1);
for J=1:JEND
  for I=1:NNPE
    SF(1,I,J)=...
        SFN(QPT(J,1),QPT(J,2),I,NNPE);
    SF(2,I,J)=...
        SFNu(QPT(J,1),QPT(J,2),I,NNPE);
    SF(3,I,J)=...
        SFNv(QPT(J,1),QPT(J,2),I,NNPE);
  end
end

%-----------------------
% SF for line quadrature
%-----------------------
        NUMQPT(2) = 3;
        QPT(1,1)= sqrt(0.6);
        QPT(1,2)=0.0;
        QPT(1,3)=-QPT(1,1);

        WT(2,1)=5.0/9.0;
        WT(2,2)=8.0/9.0;
        WT(2,3)=5.0/9.0;
```

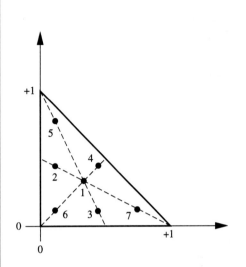

Specify weights.

Fill in *SF* matrices.

$$SF(1, I, J) = N_I(u_J, v_J)$$

$$SF(2, I, J) = \frac{\partial N_I}{\partial u}(u_J, v_J)$$

$$SF(3, I, J) = \frac{\partial N_I}{\partial v}(u_J, v_J)$$

Line quadrature:

Number of quadrature points = 3

Define coordinates.

Define weights.

```
    JEND=NUMQPT(2);
    for J=1:JEND
        for I=1:NNPS
            I1=I;
            SF(4,I,J)=SFL (QPT(1,J),I1,NNPS);
            SF(5,I,J)=SFLu(QPT(1,J),I1,NNPS);
        end
    end
```

Fill in *SF* matrices.

$$SF(4, I, J) = NS_I(u_J)$$

$$SF(5, I, J) = \frac{\partial NS_I}{\partial u}(u_J)$$

```
%   -------------------
%   Define NPSIDE array
%   -------------------
    NPSIDE(1,1)=1;
    NPSIDE(1,2)=2;
    NPSIDE(1,3)=3;
    NPSIDE(2,1)=3;
    NPSIDE(2,2)=4;
    NPSIDE(2,3)=5;
    NPSIDE(3,1)=5;
    NPSIDE(3,2)=6;
    NPSIDE(3,3)=1;
```

Create NPSIDE array.

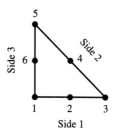

```
    else

    fprintf(1,'\n---------------------'  )
    fprintf(1,'\n Error in shafac.m    '  )
    fprintf(1,'\n NNPE =%2i',NNPE         )
    fprintf(1,' is invalid'               )
    fprintf(1,'\n---------------------\n')
    error

    end
```

Error: An element corresponding to the value of NNPE entered does not exist.

Return to main program.

```
function  f = SFN(u,v,n,NNPE)
%-----------------------------


 if NNPE == 3
     w = 1.0-u-v;
     if n == 1;
        f = u;
     elseif n == 2
        f = v;
     elseif n == 3
        f = w;
     end




 elseif NNPE == 4
     if n == 1
        f=0.25*(1.0-u)*(1.0-v);
     elseif n == 2
        f=0.25*(1.0+u)*(1.0-v);
     elseif n == 3
        f=0.25*(1.0+u)*(1.0+v);
     elseif n == 4
        f=0.25*(1.0-u)*(1.0+v);
     end




 elseif NNPE == 6
     w=1.0-u-v;
     if n == 1
        f=(2.0*u-1.0)*u;
     elseif n == 2
        f=4.0*v*u;
     elseif n == 3
        f=(2.0*v-1.0)*v;
     elseif n == 4
        f=4.0*v*w;
     elseif n == 5
        f=(2.0*w-1.0)*w;
     elseif n == 6
        f=4.0*u*A;
     end
  end
```

SFN: Function to evaluate shape functions:

Three-node element:

$N_1 = u$
$N_2 = v$
$N_3 = 1 - u - v$

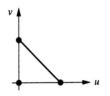

Four-node element:

$N_1 = \frac{1}{4}(1 - u)(1 - v)$
$N_2 = \frac{1}{4}(1 + u)(1 - v)$
$N_3 = \frac{1}{4}(1 + u)(1 + v)$
$N_4 = \frac{1}{4}(1 - u)(1 + v)$

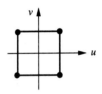

Six-node element:

$N_1 = (2u - 1)u$
$N_2 = 4vu$
$N_3 = (2v - 1)v$
$N_4 = 4vu$
$N_5 = (2w - 1)w$
$N_6 = 4uw$
where $w = 1 - u - v$

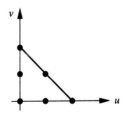

```
function f = SFNu(u,v,n,NNPE)
%----------------------------------
% Derivative of shape function with
% respect to the xi coordinate
%----------------------------------

    if NNPE == 3
        w=1.0-u-v;
        if n     == 1
            f=1.0;
        elseif n == 2
            f=0.0;
        elseif n == 3
            f=-1.0;
        end


    elseif NNPE == 4
        if n     == 1
            f=0.25*(    -1.)*(1.0-v);
        elseif n == 2
            f=0.25*(    +1.)*(1.0-v);
        elseif n == 3
            f=0.25*(    +1.)*(1.0+v);
        elseif n == 4
            f=0.25*(    -1.)*(1.0+v);
        end


    elseif NNPE == 6
        w = 1.0-u-v;
        if n     == 1
            f=4.0*u-1.0;
        elseif n == 2
            f=4.0*v;
        elseif n == 3
            f=0.0;
        elseif n == 4
            f=-4.0*v;
        elseif n == 5
            f=-4.0*w+1.0;
        elseif n == 6
            f=4.0*w-4.0*u;
        end
    end
```

SFNu: Function to evaluate $\dfrac{\partial N}{\partial u}$

Three-node element:

$N_1' = 1$
$N_2' = 0$
$N_3' = -1$

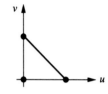

Four-node element:

$N_1' = \frac{1}{4}(-1)(1 - v)$
$N_2' = \frac{1}{4}(+1)(1 - v)$
$N_3' = \frac{1}{4}(+1)(1 + v)$
$N_4' = \frac{1}{4}(-1)(1 + v)$

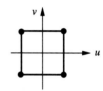

Six-node element:

$N_1' = 4u - 1$
$N_2' = 4v$
$N_3' = 0$
$N_4' = -4v$
$N_5' = -4w + 1$
$N_6' = 4w - 4u$

where $w = 1 - u - v$

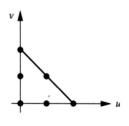

```
function f = SFNv(u,v,n,NNPE)
%-----------------------------------
% Derivative of shape function with
% respect to the eta coordinate
%-----------------------------------

    if NNPE == 3
        if n    == 1
            f=0.0;
        elseif n == 2
            f=1.0;
        elseif n == 3
            f=-1.0;
        end


    elseif NNPE == 4
        if n    == 1
            f=0.25*(1.0-u)*(   -1.0);
        elseif n == 2
            f=0.25*(1.0+u)*(   -1.0);
        elseif n == 3
            f=0.25*(1.0+u)*(   +1.0);
        elseif n == 4
            f=0.25*(1.0-u)*(   +1.0);
        end


    elseif NNPE == 6
        w=1.0-u-v;
        if n    == 1
            f=0.0;
        elseif n == 2
            f=4.0*u;
        elseif n == 3
            f=4.0*v-1.0;
        elseif n == 4
            f=4.0*w-4.0*v;
        elseif n == 5
            f=-4.0*w+1.0;
        elseif n == 6
            f=-4.0*u;
        end
    end
```

SFNv: Function to evaluate $\dfrac{\partial N}{\partial v}$

Three-node element:

$N_1' = 0$
$N_2' = 1$
$N_3' = -1$

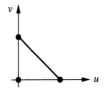

Four-node element:

$N_1' = \frac{1}{4}(1 - u)(-1)$
$N_2' = \frac{1}{4}(1 + u)(-1)$
$N_3' = \frac{1}{4}(1 + u)(+1)$
$N_4' = \frac{1}{4}(1 - u)(+1)$

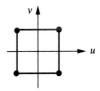

Six-node element:

$N_1' = 0$
$N_2' = 4u$
$N_3' = 4v - 1$
$N_4' = 4w - 4v$
$N_5' = -4w + 1$
$N_6' = -4u$

where $w = 1 - u - v$

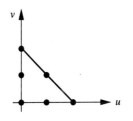

```
function f =  SFL(S,n,NNPS)
%-------------------------------
% Shape function values for lines
%-------------------------------
      if NNPS == 2
          if n      == 1
              f=-0.5*(S-1.0);
          elseif n == 2
              f= 0.5*(S+1.0);
          end
      elseif NNPS == 3
          if n      == 1
              f= -0.5*(S)*(1.-S);
          elseif n == 2
              f= (1.+S)*(1.-S);
          elseif n == 3
              f= 0.5*(1.+S)*(S);
          end
      end
```

```
function v =  SFLu(S,n,NNPS)
%---------------------------------
% Derivative of shape function wrt S
%---------------------------------
      if NNPS == 2
        if n       == 1
           f=-0.5;
        elseif n == 2
           f= 0.5;
        end
      elseif NNPS == 3
        if n       == 1;
           f = -0.5*(1.-2.*S)
        elseif n == 2
           f = -2.*S;
        elseif n == 3
           f = 0.5*(1.+2.*S);
        end
      end
```

SFL: Function to evaluate N_s

Two-node line segment:
Note: $S = u$

$N_1 = -0.5(u - 1)$
$N_2 = 0.5(u + 1)$

Three-node line segment:

$N_1 = -0.5(u)(1 - u)$
$N_2 = (1 + u)(1 - u)$
$N_3 = 0.5(1 + u)(u)$

Function to evaluate $\dfrac{dN_s}{du}$

Two-node line segment:

$N_1' = -0.5$
$N_2' = 0.5$

Three-node line segment:

$N_1' = -0.5(1 - 2u)$
$N_2' = -2u$
$N_3' = 0.5(1 + 2u)$

GAUSSIAN QUADRATURE

The necessity of evaluating definite integrals for the formulation of finite element equations requires that numerical integration schemes be incorporated into finite element codes. All numerical integrations require the evaluation of integrands at one or more points. The number of points necessary depends on the accuracy desired and the complexity of the integrand. Because the cost of an integration is directly proportional to the number of evaluations, the objective is to arrive at a scheme that, for a given degree of accuracy, requires a minimum number of evaluations. The method used in nearly all finite element codes is Gaussian quadrature.

C.1 ONE DIMENSION

In what follows, we will show that a polynomial of degree $m = 2n - 1$ can be integrated exactly by evaluating it at only n points. The coordinates of these points are not arbitrary and must be calculated using the procedure to be presented. First, however, we illustrate the concept with an elementary example.

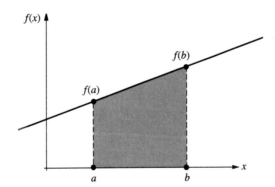

Figure C.1. Integration of a linear function.

Consider that our goal is to integrate the linear function illustrated in Fig. C.1 from $x = a$ to $x = b$. Because this is equal to the shaded area shown, it is immediately apparent that the integral can be calculated using the average of the two integrands evaluated at a and b,

$$\int_a^b f(x)\,dx = \frac{1}{2}[f(a) + f(b)](b - a) \tag{C.1}$$

or using the value of the integrand evaluated at the midpoint, $(a + b)/2$:

$$\int_a^b f(x)\,dx = f\left(\frac{a + b}{2}\right)(b - a) \tag{C.2}$$

Obviously, the latter is preferable because it requires only one evaluation of the integrand.

315

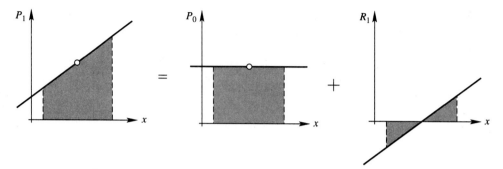

Figure C.2.   Factorization of a linear function.

We can interpret this procedure as shown in Fig. C.2, where a first-degree polynomial, P_1, is separated into two polynomials—one of zeroth degree, P_0, whose constant value is equal to P_1 evaluated at $(a + b)/2$, and one of first degree, R_1, equal to $P_1 - P_0$. We now write

$$\int_a^b P_1 \, dx = \int_a^b P_0 \, dx + \int_a^b R_1 \, dx \tag{C.3}$$

where, because of our previous analysis, we know

$$\int_a^b P_1 \, dx = \int_a^b P_0 \, dx \tag{C.4}$$

This means, of course,

$$\int_a^b R_1 \, dx = 0 \tag{C.5}$$

We have thus split our original polynomial into two polynomials, one of lower degree (degree = 0) and another of the same degree (degree = 1) but whose integral is zero. Thus, we were able to evaluate the integral of our original polynomial by evaluating the integral of a lower-degree polynomial. In this example it was necessary to know two things: the location at which to evaluate P_1 (midpoint) and what to do after it was evaluated (multiply it by $b - a$). The location is the Gaussian quadrature point, and the multiplication term is the Gaussian weight.

We now show that any polynomial of degree $2n - 1$ can be split into two polynomials, P_n and R_{2n-1}, so that

$$\int_a^b P_{2n-1} \, dx = \int_a^b P_n \, dx + \int_a^b R_{2n-1} \, dx = \int_a^b P_n \, dx \tag{C.6}$$

because

$$\int_a^b R_{2n-1} \, dx = 0 \tag{C.7}$$

Both P_n and R_{2n-1} will be defined by values of P_{2n-1} at specified quadrature points. However, only P_n will be of importance because the integral of R_{2n-1} will always be zero.

The n quadrature points for a polynomial of degree $2n - 1$ are always given with respect to the normalized limits of integration $a = -1$ and $b = +1$. Once the points within these limits are known, their locations between any other limits are found by using simple proportionality. We develop the general procedure for determining these points using $n = 3$; the extension to a larger number of points will be obvious. Hence, we seek three points, a, b, and c, so that if they are roots of any fifth-degree polynomial, then that polynomial will have a zero integral between -1 and $+1$. The most general fifth-degree polynomial that has roots at these points can be written as

$$y(x) = \left(C_0 + C_1 x + C_2 x^2\right)\left[(x - a)(x - b)(x - c)\right] \tag{C.8}$$

Our goal is to make the integral of this function, between $x = -1$ and $x = +1$, equal to zero for all values of C_0, C_1, and C_2. This requires that the following three equations be satisfied:

$$\int_{-1}^{+1} (x - a)(x - b)(x - c)\,dx = 0$$

$$\int_{-1}^{+1} x(x - a)(x - b)(x - c)\,dx = 0 \tag{C.9}$$

$$\int_{-1}^{+1} x^2(x - a)(x - b)(x - c)\,dx = 0$$

which gives us, after integration and simplification,

$$-(a + b + c) + 3abc = 0$$

$$3 + 5(ab + bc + ca) = 0 \tag{C.10}$$

$$-3(a + b + c) + 5abc = 0$$

The solution to this set of nonlinear algebraic equations is

$$a = -\sqrt{0.6}$$

$$b = 0 \tag{C.11}$$

$$c = +\sqrt{0.6}$$

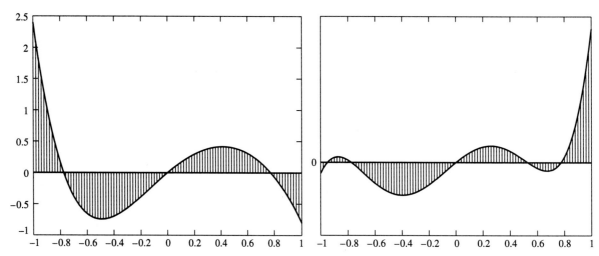

Figure C.3.   Fifth-degree polynomials with Gaussian roots.

Figure C.3 shows two fifth-degree polynomials, both having roots at the three Gaussian quadrature points just found. In each case, the shaded area equals zero.

We now show how we will use this information to evaluate the integral of a general fifth-degree polynomial. Consider the curves illustrated in Fig. C.4 .

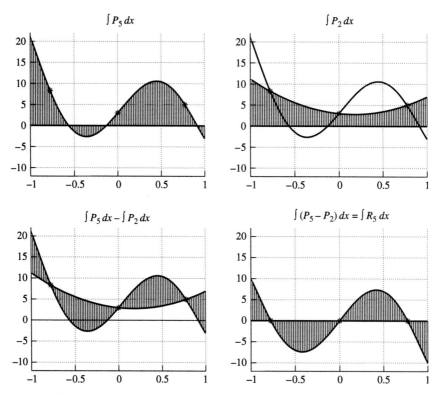

Figure C.4.   Integration of a fith-degree polynomial.

The integral P_5 represents our general polynomial, and its values at the three quadrature points are marked. P_2 is the unique second-degree polynomial that passes through these same three points. Its integral, $\int P_2 \, dx$, is the same as that of P_5, that is, the two shaded areas are equal. This is true because the area indicated as $\int P_5 \, dx - \int P_2 \, dx$ is equal to zero. It is the same area shown as $\int (P_5 - P_2) \, dx = \int R_5 \, dx$, and because R_5 is a fifth-degree polynomial having roots equal to the Gaussian quadrature points, its integral, as well as that of any other fifth-degree polynomial with these same roots, is equal to zero.

We have only to evaluate $\int P_2 \, dx$, using the three values of P_5 that P_2 shares. This will be done using three weights associated with each quadrature point, such that

$$\int_{-1}^{+1} P_2 \, dx = W_1 P_5(x_1) + W_2 P_5(x_2) + W_3 P_5(x_3) \tag{C.12}$$

These weights are easily evaluated by noting that the above equation must give exact solutions to any polynomial of degree 5 or less. In particular, it must integrate a constant, a linear function, and a second-degree polynomial exactly. Thus,

$$\int_{-1}^{+1} 1 \, dx = 2 \qquad = W_1 1 + W_2 1 + W_3 1$$

$$\int_{-1}^{+1} x \, dx = 0 \qquad = W_1 x_1 + W_2 x_2 + W_3 x_3 \tag{C.13}$$

$$\int_{-1}^{+1} x^2 \, dx = 2/3 \qquad = W_1 x_1^2 + W_2 x_2^2 + W_3 x_3^2$$

This gives us three linear equations for our three unknown weights. In matrix form,

$$\begin{bmatrix} 1 & 1 & 1 \\ -\sqrt{0.6} & 0 & \sqrt{0.6} \\ 0.6 & 0 & 0.6 \end{bmatrix} \begin{Bmatrix} W_1 \\ W_2 \\ W_3 \end{Bmatrix} = \begin{Bmatrix} 2 \\ 0 \\ 2/3 \end{Bmatrix} \tag{C.14}$$

The solution is

$$\begin{Bmatrix} W_1 \\ W_2 \\ W_3 \end{Bmatrix} = \begin{Bmatrix} 5/9 \\ 8/9 \\ 5/9 \end{Bmatrix} \tag{C.15}$$

Note that the first equation of this set, regardless of the number of points, states that the sum of the weights equals 2.

The extension to more points can now be made. The general form for Eq. C.8 is

$$y(x) = \left(C_0 + C_1 x + \cdots + C_n x^n \right) \left[(x - x_1)(x - x_2) \cdots (x - x_n) \right] \tag{C.16}$$

This is the most general polynomial of degree $2n - 1$ that has n roots, $x_1 - x_n$. If the integral of this polynomial is to be zero for all values of $C_0 - C_n$, then we have n equations for the n unknown roots,

$$\int_{-1}^{+1} x^{n-i} \left[(x - x_1)(x - x_2) \cdots (x - x_n) \right] dx = 0 \tag{C.17}$$

where i ranges from 1 to n. The equations to determine the corresponding weights are those represented by Eq. C.13, but expanded to represent n linear equations.

Equation C.17 represents a set of highly nonlinear equations for the quadrature coordinates. A simplification can be made by noting the coordinates are always symmetrical about $x = 0$. For an even number of points, this reduces the number of unknowns by half. For an odd number of points, it requires that the midpoint be $x = 0$ and reduces the remaining number of unknowns by half. However, even with this reduction, the solution of the equations becomes difficult as the number of points is increased. Another approach is to write Eq. C.16 as

$$y(x) = \left(C_0 + C_1 x + \cdots + C_n x^n \right) \left[B_0 + B_1 x + \cdots + B_n x^n \right] \tag{C.18}$$

that is, in terms of the coefficients of P_n rather than its roots. In this form, we have a set of linear equations for the coefficients, B_i. Once these coefficients have been determined, the problem is reduced to one of finding the roots of a given polynomial. There are well-established numerical methods for doing this, and they will not be discussed here. It is important, however, to note that the resulting polynomial found in this fashion is, in fact, the Legendre polynomial of the same degree. Hence, the quadrature points we seek are the roots of the corresponding Legendre polynomial. It is for this reason that the method is often referred to as Gauss-Legendre quadrature.

C.2 QUADRATURE FOR TWO DIMENSIONS

The extension of one-dimensional quadrature to two-dimensional quadrature in a rectangular domain is straightforward. The normalized coordinates for the rectangle are shown in Fig. C.5 .

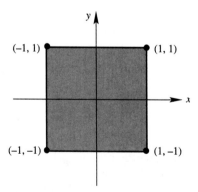

Figure C.5. Unit square for Gaussian quadrature.

Let $\phi(x, y)$ be the integrand that we wish to integrate over the area shown, and let n be the number of quadrature points in one direction as defined in the previous section. Then

$$\int_A \phi(x, y)\, dA = \int_{-1}^{+1} \int_{-1}^{+1} \phi(x, y)\, dx\, dy$$

$$\int_{-1}^{+1} \int_{-1}^{+1} \phi(x, y)\, dx\, dy = \int_{-1}^{+1} \left[\int_{-1}^{+1} \phi(x, y)\, dx \right] dy$$

$$\int_{-1}^{+1} \left[\int_{-1}^{+1} \phi(x, y)\, dx \right] dy = \int_{-1}^{+1} \left[\sum_{i=1}^{n} W_i \phi(x_i, y) \right] dy$$

$$\int_{-1}^{+1} \left[\sum_{i=1}^{n} W_i \phi(x_i, y) \right] dy = \sum_{j=1}^{n} W_j \left[\sum_{i=1}^{n} W_i \phi(x_i, y_j) \right]$$

Hence,

$$\int_A \phi(x, y)\, dA = \sum_{j=1}^{n} \sum_{i=1}^{n} W_i W_j \phi(x_i, y_j)$$

For triangular regions the method is similar but not as straightforward. For those interested, the procedure is given in [Cowper 1973]. Only the seven-point quadrature shown below is used in the codes in this text. Additional coordinates and weights can be obtained from the reference.

Point	x	y	Weight	
1	A_0	A_0	W_0	$A_0 = 1/3$
2	A_1	B_1	W_1	$A_1 = 0.059715871789770$
3	B_1	A_1	W_1	$B_1 = 0.470142064105115$
4	B_1	B_1	W_1	$A_2 = 0.797426985353087$
5	B_2	A_2	W_2	$B_2 = 0.101286507323456$
6	B_2	B_2	W_2	$W_0 = 0.1125$
7	A_2	B_2	W_2	$W_1 = 0.066197076394253$
				$W_2 = 0.062969590272413$

C.2.1 Area Coordinates. The preceding coordinates for quadrature within a triangle are usually given in terms of area coordinates. These coordinates have the advantage that they locate points within a triangular region irrespective of the shape of the triangle. Consider the triangle shown in Fig. C.6 and the particular point shown by the dark circle. Three areas within the triangle are defined by that point: A_1, A_2, and A_3. What is more important is that these three areas uniquely define the location of the point. That is, no other point can create areas with the same magnitudes. Thus, these values represent valid coordinates for locating points within a triangle. When these areas are divided by the total area of the triangle, they are called *area coordinates*. Note that because their sum must be equal to unity, they are not independent. That is, only two of the three

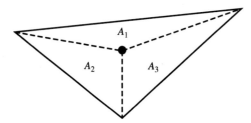

Figure C.6.   Area coordinates.

area coordinates are necessary to define a point within a triangle. However, it is customary and convenient to define all three values. There are many interesting properties associated with the use of area coordinates. As an example, consider the loci of points created by holding any one of the coordinates to a constant value. This would be a straight line parallel to one side of the triangle.

Although there are many advantages to using area coordinates, we have preferred to think in terms rectangular coordinates. Interestingly, the coordinates x and y shown in Fig. C.7 for any given point correspond exactly to two of the area coordinates for that point. Hence, quadrature points given in terms of area coordinates can be interpreted as coordinates given in terms of the x, y coordinates shown. This is the way we have interpreted them.

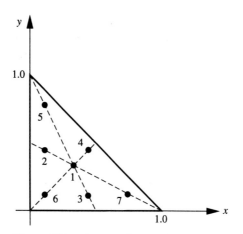

Figure C.7.   Seven-point quadrature.

D

AUXILIARY CODES

Finite element analyses require the processing of large data files, which is expedited by using auxiliary codes referred to as *preprocessors* and *postprocessors*. The preprocessors create the data associated with the mesh as well as any data associated with creating more efficient storage and run times. The postprocessors are used to help analyze the output data. Most often the latter codes are used for plotting the results for graphical interpretation. Such codes, both pre- and postprocessing codes, can be run independently of the finite element code; hence, the user is free to select from a variety of commercially available codes.

Five such codes are included with this book:

mesh.m	A preprocessor that generates the XORD, YORD, NPcode, and NP arrays associated with a given mesh.
newnum.m	A preprocessor to create the NWLD array for bandwidth reduction.
topo.m	A postprocessor for plotting contours and color mappings of the results.
squash.m	A postprocessor for elastic.m that can be used to plot the deformed shape of the solid or to plot the displacement vectors at each node.

Although we will not give a listing of these codes here, they are available on the text's Web page and are well documented to assist those interested in understanding their logic. In what follows, we describe the purpose of each code and how to use it.

D.1 PROGRAM mesh.m

Program mesh.m is a mesh generator that prepares input for finite element programs. Three-, four-, six- and eight-node elements can be specified. The program generates meshes by piecing together "quadrilateral" subsections (called LOOPs) that have parabolic sides. The user specifies the shape of these LOOPs by designating eight coordinates for each LOOP.

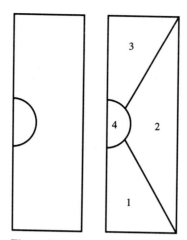

Figure D.1.   Region to be analyzed.

As an example, consider the geometry shown on the left in Fig. D.1. This region was associated with a finite element analysis for which the region around the half circle was important; hence, it was desired to have a finer mesh in that region than elsewhere. The region was divided into the four LOOPs shown on the right in Fig. D.1. LOOPs 1, 2, and 3 are easily recognized as "quadrilaterals," each with three straight sides and one parabolic side. The three parabolic sides join the fourth LOOP to form three of its sides, making an almost circular shape. After the LOOPs were defined, the number of elements for each LOOP was selected to produce the mesh shown in Fig. D.2. How to define the LOOP geometry, specify the number of elements, and piece together the LOOPs will now be explained.

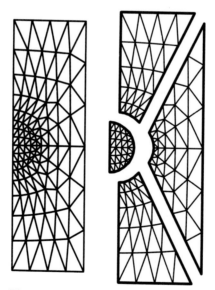

Figure D.2.   Mesh with LOOPs.

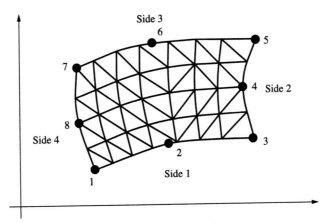

Figure D.3. LOOP nodes and side numbering.

Defining the Boundaries (Sides). The LOOP boundaries are defined by specifying the x and y coordinates of the eight LOOP nodes shown in Fig. D.3. The LOOP nodes must always be numbered counterclockwise, with node 1 at one of the four vertices of the quadrilateral. Each side is uniquely defined by the parabola passing through its three nodes. The "mid"-side node is not only used to define the parabola, but it can also be used to proportion the element density within the LOOP. This use of a mid-side node can be seen in Fig. D.2 where elements in LOOPs 1, 2, and 3 are pushed closer to the half-circle region. Sides are numbered counterclockwise starting with the side containing LOOP nodes 1, 2, and 3. For each LOOP, coordinates of eight LOOP nodes must be entered into program mesh.m.

Designating the Number of Elements in a LOOP. The number of elements in a LOOP is designated indirectly by specifying the number of divisions along sides 1 and 2, which will also designate the divisions on sides 3 and 4. Thus, in Fig. D.2 we see that the bottom and top sides of LOOP 1 were designated as having four divisions while the other two sides were designated as having five divisions. This produced 4×5 or 20 subquadrilaterals, each with two elements. mesh.m assumes that each LOOP is thus divided; therefore, side 3 will have the same number of divisions as side 1, and side 4 will have the same number of divisions as side 2, and only the numbers of divisions along sides 1 and 2 are entered as data.

Joining LOOPs. A complete mesh is formed by piecing together LOOPs. This piecing is done sequentially: thus, the first LOOP formed is the foundation LOOP, with subsequent LOOPs joined either to it or to other LOOPs that have already been defined. As each LOOP is defined, the user must specify for each of the four sides of the current LOOP

1. What previous LOOP, if any, it should be joined to
2. To which side of the previous LOOP it should be joined

This information is supplied with one input record that lists two numbers for each of the four sides of the current LOOP. The first number represents a previous LOOP and the second number indicates the side of the previous LOOP to which the current LOOP side is to be joined. Thus, the record

side 1		side 2		side 3		side 4		Current LOOP
a	*b*	*c*	*d*	*e*	*f*	*g*	*h*	
elem.	side	elem.	side	elem.	side	elem.	side	Adjacent LOOPs

indicates that side 1 of the current LOOP is to be joined to side *b* of LOOP *a*. Side 2 of the current LOOP is to be joined to side *d* of LOOP *c*, and so forth. If a side is not to be joined to a previously formed LOOP, the two numbers should be specified as zero.

When joining two LOOPs, it is essential that the two sides to be joined have the same number of divisions. If this is not the case, mesh.m gives an error message and aborts. Normally, it is best to specify the LOOP nodes associated with the two sides that are to be joined to have the same coordinates; if not, the element nodes along the joined side will have coordinates dictated by the LOOP nodes of the first LOOP. This can result in significant distortion of the elements in the current LOOP.

This specification, the joining of LOOPs, is important because it indicates that the nodes on the joined side of the new LOOP have the same global nodal numbers as those on the side of the previous LOOP. If this is not specified, there will be two numbers specified for each of the nodes shown along the common boundary. The adjacent elements, which appear to have a common side, will in fact be independent from each other. Thus, the LOOP boundary will represent a line of discontinuity of the primary variable. This, by the way, may be exactly what is specified by the problem, in which case it is necessary to specify that the LOOPs are not joined.

Nodal Point and Element Numbering. Program mesh.m assigns numbers to the nodes and the elements as each LOOP is formed. Therefore, the numbering scheme is dependent on the LOOP order. The first LOOP will contain the first element number and the first node number; the last LOOP will contain the last node number and the last element number. Within any LOOP, the numbering for the nodes begins at LOOP node 1 and increases along side 1. When all the nodes along side 1 have been numbered, the numbering continues along the next row of nodes, and so forth. The numbering of the elements follows much the same pattern. Figure D.4 illustrates how the numbers would be ordered for a typical LOOP.

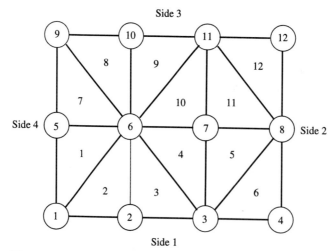

Figure D.4. Order of node and element numbering within a LOOP.

NPCODE.m, *a User INCLUDE Code.* Before mesh.m creates the output data files, a user INCLUDE file is called on to assign NPcode values to each node. These values have meaning only to the user and are not used in any of the finite element codes. However, the user can use these numbers in any other INCLUDE code. They are used most often in the INITIAL.m codes that set boundary and initial conditions for a particular analysis.

Two arrays are available in mesh.m that are particularly helpful for assigning NPcode values. These are NDIV(I,J), which specifies the number of divisions on side *J* of LOOP *I*, and LNP(I,J,K), which specifies the global node number of the *K*th node on side *J* of LOOP *I*. Here *K* represents the node number based on a counter clockwise numbering, starting with the first node on side *J*.

Output. At the conclusion of mesh.m, the user will have the following ASCII files:

MESHo
NODES
NP

These files are the same as those used in the FEM programs described in the text.

Some Final Hints and Cautions

1. When you are designing a complex mesh and you have a problem you cannot understand, then run mesh.m with only the first LOOP. If that looks fine, then run it with the first two LOOPs. Continue in this way until you find your problem. Sometimes a particular problem can best be found by running each LOOP as a separate mesh. All of this is easily done by editing the DATAM file, defined in the next section.
2. During the design stage, use only one or two divisions per side. After you obtain a mesh that represents the geometry desired, you can then increase the number of divisions for your final run.
3. Remember, LOOPs can only be joined to previously defined LOOPs. Program mesh.m cannot look ahead, only backward.
4. Do not try to push the mid-side LOOP coordinates more than 25% off the center position. If you need a more drastic gradation of spacing, simply divide the LOOP into two LOOPs.
5. The file NPCODE.m is included in mesh.m immediately before the output is saved. The user therefore has a chance to change, add to, or subtract from the data generated by mesh.m before the output is saved. This can be most useful when, for example, one or more nodal coordinates must be slightly altered in order to match the geometry of the actual problem more exactly, or when one or more elements or nodes must be added to create a particular geometry not possible to create with LOOPs. In fact, it is possible to let mesh.m create a simple rectangular region, and then in NPCODE.m write a few lines of code that will map the rectangle into a ring or some other desired shape.

A Complete Example for Using mesh.m. As an example, the data files and NPCODE.m INCLUDE file to create the mesh shown in Fig. D.2 are now given. The LOOP numbers are those shown in Fig. D.1.

Data File: LOOPs

```
%----------------------
%    NUMLPS      NNPE
%----------------------
          4        4
%----------------------
%    NDIV-1     NDIV-2
%----------------------
          4        5
          5        6
          4        5
          6        4
```

Mesh is made up of four LOOPs.
Four-node elements are requested.

Divisions:
For example: second row refers to second LOOP.
There will be five divisions on side 1,
six divisions on side 2.

Data File: JOIN

```
%     JOIN ARRAY
%------------------------
     0 0   0 0   0 0   0 0
     1 2   0 0   0 0   0 0
     0 0   2 3   0 0   0 0
     0 0   1 3   2 4   3 1
%------------------------
```

There are four lines in this file, one for each LOOP. Each group of two integers refers to a given side of the LOOP for its line; hence, there are four groups of two integers.

Example:
The third line is for LOOP 3. It states that its sides 1, 3, and 4 are not connected to any *previous* LOOP. However, its side 2 (second pair of numbers) is joined to side 3 of LOOP 2.

Data File: COORD

```
%     LOOP Coordinates (8 per loop)
%-------------------
      0        -16
      5        -16
     10        -16
      5.29      -7.30
      2.34      -1.87
      1.30      -2.70
      0        -3
      0        -7.5
%-------------------
      2.34      -1.87
      5.29      -7.30
     10        -16
     10         0
     10        16
      5.29      7.30
      2.34      1.87
      3         0
%-------------------
      0         3
      1.30      2.70
      2.34      1.87
      5.29      7.30
     10        16
      5        16
      0        16
      0         7.5
%-------------------
      0         3
      0         0
      0        -3
      1.30     -2.70
      2.34     -1.87
      3         0
      2.34      1.87
      1.30      2.70
%-------------------
```

There are four sets of eight coordinates, one set for each LOOP.

The first column in each set are the *x* coordinates, and the second column are the *y* coordinates.

The coordinates are listed in counterclockwise order around the LOOP, starting at any LOOP corner (i.e., not a mid-side point).

The first three nodes listed define side 1 of the LOOP. The other three sides are numbered counterclockwise.

NPCODE.m INCLUDE Code

```
%--------------------------------
% NPCODE.m
%
% A user INCLUDE code
%
% LNP(I,J,K) = node K, on side J
% of element I.
%--------------------------------

% --------------------------------
% Set n-factor for number of nodes
% on a side
% --------------------------------
   if NNPE == 6 | NNPE == 8
      n=2;
   else
      n=1;
   end



% --------------------------------
% Set NPcode = 1 on lower boundary
% --------------------------------
   IEND = n*NDIV(1,1)+1;
   for I=1:IEND
      NI=LNP(1,1,I);
      NPcode(NI)=1;
   end

% --------------------------------
% Set NPcode = 2 on right side of mesh
% --------------------------------
   IEND = n*NDIV(2,2)+1;
   for I=1:IEND
      NI=LNP(2,2,I);
      NPcode(NI)=2;
   end

% --------------------------------
% Set NPcode = 3 on top boundary
% --------------------------------
   IEND = n*NDIV(3,3)+1;
   for I=1:IEND
      NI=LNP(3,3,I);
      NPcode(NI)=3;
   end
```

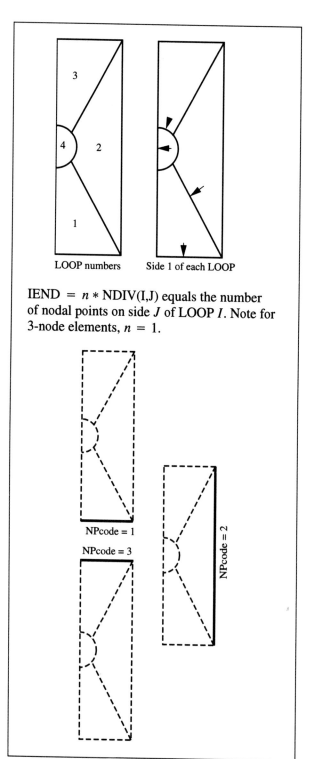

LOOP numbers   Side 1 of each LOOP

$IEND = n * NDIV(I,J)$ equals the number of nodal points on side J of LOOP I. Note for 3-node elements, $n = 1$.

NPcode = 1

NPcode = 3

NPcode = 2

```
% --------------------------------------
% Set NPcode = 4 on left side of mesh
% --------------------------------------
   IEND = n*NDIV(1,4)+1;
   for I=1:IEND
      NI=LNP(1,4,I);
      NPcode(NI)=4;
   end


   IEND = n*NDIV(4,4)+1;
   for I=1:IEND
      NI=LNP(4,4,I);
      NPcode(NI)=4;
   end


   IEND = n*NDIV(3,4)+1;
   for I=1:IEND
      NI=LNP(3,4,I);
      NPcode(NI)=4;
   end




% --------------------------------
% Set NPcode for corner points
% --------------------------------
   NI=LNP(1,1,1);
   NPcode(NI) = 11;

   NI=LNP(1,2,1);
   NPcode(NI) = 12;

   NI=LNP(3,3,1);
   NPcode(NI) = 22;

   NI=LNP(3,4,1);
   NPcode(NI) = 21;
```

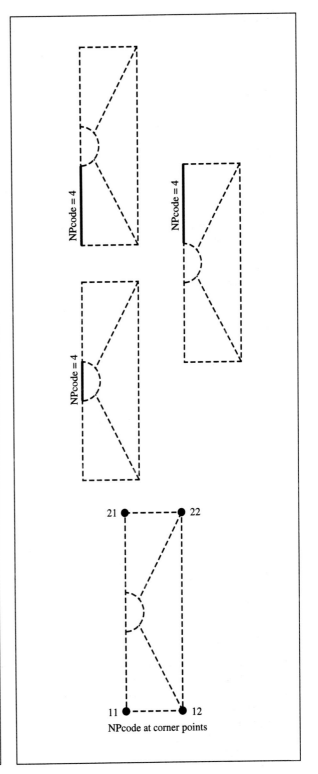

NPcode at corner points

D.2 PROGRAM newnum.m FOR BANDWIDTH REDUCTION

Before trying to understand bandwidth reduction schemes, it is necessary to understand the relationship between bandwidth and nodal numbering. Finite element equations are related to each other only through common elements. Therefore, the goal of any bandwidth reduction program is to have nodes that are connected by common elements be as close in numerical value as possible.

The connectivity of nodes through elements is designated in the NP-array. For that reason, this is the only data necessary for a bandwidth reduction program. The code described here begins with the user specifying one or more nodes (using the original numbers) as starting nodes—those that will be the first numbers in the new numbering scheme. For example, if five nodes are designated, then their new nodal numbers will be 1 through 5. The line connecting these nodes is referred to as the first wave of nodes. The second wave consists of all nodes that are linked to nodes in the first wave through common elements. The nodes in the second wave are then given the next consecutive numbers in the new order. This process continues until all nodes have been given new numbers. In the end, all elements have node numbers that differ by no more than the number of nodes in the longest two consecutive waves, and often less. Hence, the goal is to create a set of waves moving through the mesh so as to reduce the length of the longest wave. This may take several tries using different initial waves.

As an example, consider the sequence of waves shown in Fig. D.5a. In this case, the nodes shown as wave 1 were specified as the starting nodes. These nodes were therefore assigned the numbers 1, 2, 3 and 4. The next wave are those nodes shown as wave 2, and so forth.

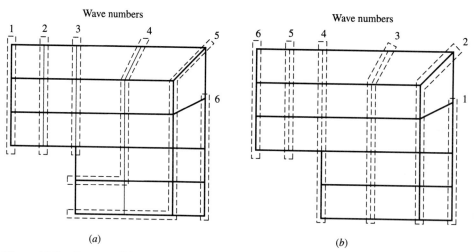

Figure D.5.   Two possible waves.

Notice that wave 5 has eight nodes; hence, we know that a bandwidth greater than 8 will be required. Had the user designated the first wave of nodes as those shown in Fig. D.5b, then the maximum number of nodes in any wave would have been 6. The selection of nodes to appear in the first wave is therefore important.

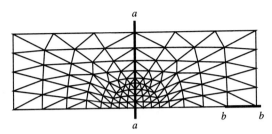

Figure D.6.   Tracing a line of nodes through a mesh.

You can often determine this by first looking at the "thickest" part of the mesh (that region containing the largest number of nodes) and locating a line that cuts through it with as few nodes on it as possible. Consider the mesh of Fig. D.6. The thickest region is in the center of the figure, and the line *a-a* shown has only 10 nodes. After locating such a line, draw a line through all nodes that are connected to nodes on this line. Follow that with a similar line through the nodes connected to the second line, and so forth. If you have done this using the right half of the figure, the line *b-b* should be your final line. The two nodes on this line, therefore, would be good choices to specify as the starting wave. Unfortunately, starting with these two nodes does not lead to line *a-a*; the process is not always reversible. However, it is easily done and most often produces a fairly good starting wave. You might find that several starting waves are necessary to result in a near-optimal numbering system.

Program newnum.m. The procedure just presented is used by program newnum.m to produce a new nodal numbering for any mesh made up of any element type. The MESHo data file is needed to define the type of element in the mesh and the number of elements and nodes. Data file NP must also be available to establish nodal connectivity. The user is asked to type in the nodes to be used in the first wave. Note that these are entered in terms of their old numbers—the ones used in the NP file. The program then computes the new nodal numbers and reports the results to the user. As an example, if you entered the starting nodes on line *b-b* for the sample mesh, the following would appear on your screen:

```
-------------------------------------------
                  Bandwidth
-------------------------------------------
              IB =     13

    Symmetric:    Bandwith = VPN*IB
    NonSymmetric: Bandwith = 2*VPN*IB-1

    VPN = variables per node
-------------------------------------------
```

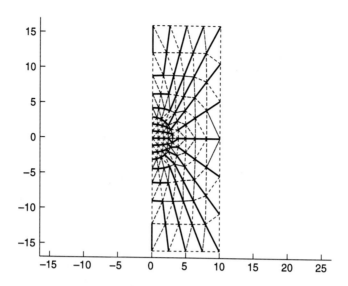

The plot shows each wave created during the analysis. A careful inspection of these waves often reveals how a better starting wave might be selected. Once again, it may be necessary to try three or more starting waves to arrive at a near-optimal bandwidth. In addition to the preceding data, the file NWLD will be placed in your working directory. This file contains the new nodal numbers in the order of the old numbers:

$$\text{New number} = \text{NWLD(old number)}$$

The last entry in this, file NWLD(NUMNP+1), is the bandwidth created by the new numbering scheme. Notice that this bandwidth is for a symmetric matrix with one variable per node. Bandwidths for other conditions can be ascertained by the equations given.

D.3 PROGRAM topo.m

Program topo.m plots contours of constant Φ values and/or color mappings of the values. It is an interactive program that asks the user what type of plot is desired. The dependent variable to be plotted must be in the file PHI, with all mesh data in the same directory.

After the user types topo, the following questions will be asked:

```
ENTER:
--------------------------------
y if you wish mesh to be plotted
--------------------------------
< n

ENTER:
--------------------------------------
do you wish countours to be plotted
   0   if you do not
   n   for n-numbers of contours
--------------------------------------
< 15

ENTER:
--------------------------------------
y if you wish a color mapping
--------------------------------------
< n

ENTER:
--------------------------------------
   0   for no SYMMETRY
   1   for SYMMETRY about X axis
   2   for SYMMETRY about Y axis
   3   for SYMMETRY about both axes
--------------------------------------
< 0

ENTER:
---------
TITLE
---------
< Test Problem
```

The following figure was created using the contour option.

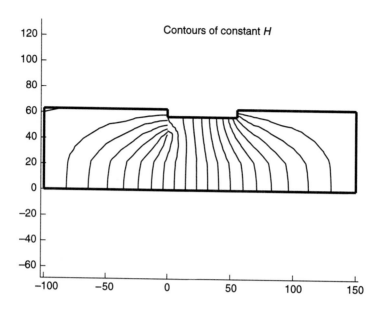

The symmetry specifications have to do only with plotting. If you have run an analysis using symmetry about the y axis, and you wish the plot to be displayed as if you ran the analysis using both sides, then specify that you want symmetry about the y axis. If you want only the half you actually used for the analysis, then specify no symmetry.

Program topo.m creates all plots based on a three-node triangular region. Higher-order elements are subdivided in the following manner:

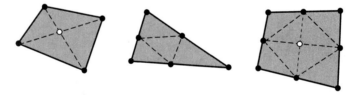

The open circles are nodes added only for plotting, and the Φ value used at each of these points is determined by averaging the surrounding nodes.

Any combination of options can be used. For example, you can specify that the mesh be shown with contours superimposed on color. Also, you can use topo.m simply to plot the mesh.

D.4 PROGRAM squash.m

Program squash.m is an interactive program for the visualization of the results from elastic.m. It can be used to plot the deformed (squashed) mesh or plot the nodal displacement vectors. The following examples illustrate these two applications.

```
ENTER:
---------------------
displacement scale
---------------------
< 1

ENTER:
-------------------------------
   1  for displaced mesh
   2  for displacement vectors
-------------------------------
< 1

ENTER:
----------------------------------
   0  for no SYMMETRY
   1  for SYMMETRY about X axis
   2  for SYMMETRY about Y axis
   3  for SYMMETRY about both axes
----------------------------------
< 3
```

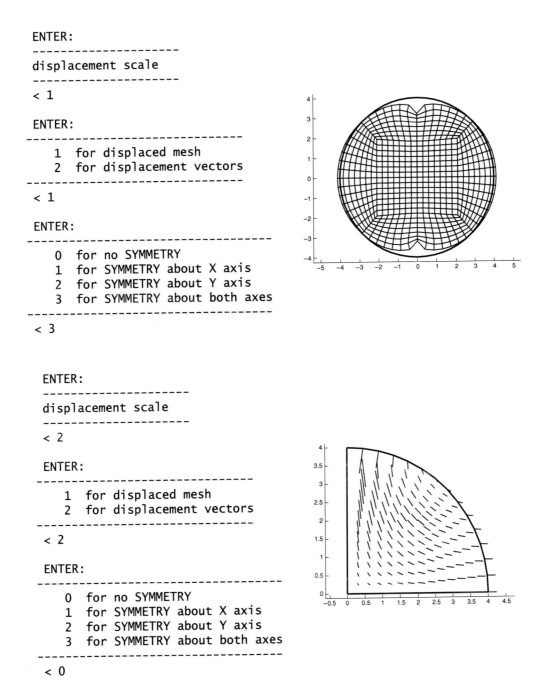

```
ENTER:
---------------------
displacement scale
---------------------
< 2

ENTER:
-------------------------------
   1  for displaced mesh
   2  for displacement vectors
-------------------------------
< 2

ENTER:
----------------------------------
   0  for no SYMMETRY
   1  for SYMMETRY about X axis
   2  for SYMMETRY about Y axis
   3  for SYMMETRY about both axes
----------------------------------
< 0
```

All mesh data and the SOLUTION file from elastic.m must be in the working directory.

Note that this is from an analysis of a Brazilian tensile test, and the actual FEM analysis was conducted on only the first quadrant of the cross-section. Hence, in the first example, symmetry about both axes was requested in order to produce the complete cross-section. However, in the second example, no symmetry was requested, and only the actual region of analysis was plotted.

Because the actual deformations are usually small, a scale factor is used to exaggerate the deformations and the deformation vectors. This scale factor is a multiple of

$$(0.1)*(XMAX-XMIN+YMAX-YMIN)/Dmax;$$

where Dmax is the maximum displacement of any node. Hence, a scale factor of 1 will plot the maximum deformation as being between 0.1 and 0.2 times the maximum size of the mesh. Depending on the particular problem, the user may want to visualize a larger deformation or a smaller deformation by using other values of the scale factor. The program is very fast, so it is not time consuming to try several scale factors to arrive at the desired effect.

Although this program is intended primarily to illustrate deformation, it is a valuable tool for debugging a problem. Both the deformed mesh and displacement vectors can be used to visually detect errors associated with boundary conditions that were not been properly specified.

Note that the preceding figures are not those that will appear on your screen. The original mesh will appear on your screen in light green. However, for clarity, this has been left out of the figures. You may want to edit squash.m to make plotting the original mesh a choice to be made by the user when the program is run.

BIBLIOGRAPHY

Bathe, Klaus-Jürgen. *Finite Element Procedures*. Prentice-Hall, Englewood Cliffs, N.J., 1996.
A comprehensive work in one volume. It presents mathematical theory, practical procedures, examples, FOR-TRAN codes, and extensive references.

Becker, Eric B., Graham F. Carey, and J. Tinsley Oden. *Finite Elements: An Introduction, Volume I*. Prentice-Hall, Englewood Cliffs, N. J., 1982.
This is the first volume in a series of excellent books on the finite element method. The mathematics is rigorous yet within the grasp of graduate students in engineering.

Crandall, Stephen H. *Engineering Analysis: A Survey of Numerical Procedures*. McGraw-Hill, New York, 1956.
This book is a classic and should be in the library of all engineers who are interested in numerical methods. Although published before the rise of the finite element method, it thoroughly covers the method of weighted residuals.

Cowper, G. R. "Gaussian Quadrature Formulas for Triangles." *International Journal for Numerical Methods in Engineering* **7**, 1973, pp. 405–408.
The derivation of the data for triangular quadrature presented in Chapter 9 and in Appendix C is in this paper.

Frocht, Max Mark. *Photoelasticity, Volume I*. John Wiley & Sons, New York, 1941 (8th printing 1966).
The photoelastic figures in Chapter 11 were taken from this book, which covers the topic in much detail.

Hughes, Thomas J. R. *The Finite Element Method: Linear Static and Dynamic Finite Element Analysis*. Dover, Mineola, N.Y., 1987, 2000.
An excellent and comprehensive work. Recommended for those who wish to cover the material in more depth.

Irons, Bruce M. "Engineering Applications of Numerical Integration in Stiffness Methods." *AIAA Journal*, Technical Notes, November 1966.
and
"Numerical Integration Applied to Finite Elemen Methods," Working Session No. 4, Paper No. 19, in *International Symposium: The Use of Electronic Digital Computers in Structural Engineering*. University of Newcastle upon Tyne, 1966.
The introduction of isoparametric elements.

Wilson, Edward Lawrence. *The Finite Element Analysis of Two-Dimensional Structures*. University of California, Berkeley, 1963. (Available through University Microfilms, Ann Arbor, Michigan.)
This short dissertation might have done more to launch the finite element revolution than any other document. Wilson's concise code, listed as an appendix, must have been typed into hundreds of keypunch machines throughout the engineering world. Today, many codes, including those in this book, reflect the logic and notation found in his code.

Zienkiewicz, O. C., and R. L. Taylor. *The Finite Element Method, Volumes 1–3*. Butterworth-Heinemann, London, 2000.
A three-volume set: (1) Basis, (2) Solid Mechanics, and (3) Fluid Mechanics. This set covers the field. Highly recommended.

INDEX